Theories and Practices of Development

Global economic crisis and the implications of global environmental change have led academics and policy-makers to consider how 'development' in all parts of the world should be achieved. However, 'development' has always been a contested idea. While often presented as a positive process to improve people's lives, the potential negative dimensions of 'development' for people and environments must also be recognized.

Theories and Practices of Development provides a clear and user-friendly introduction to the complex debates around how development has been understood and achieved. The second edition has been fully updated and expanded to reflect global political and economic shifts, as well as new approaches to development. The rise of China and India is given particular attention, as is the global economic crisis and its implications for development theories and practice. There are new sections on faith-based development, and the development dimensions of climate change, as well as greater engagement with development theories as they are put into practice in the Global North.

The book deals with the evolution of development ideas and policies, focusing on economic, political, social, environmental and spatial dimensions. It highlights how development cannot be considered as a neutral concept, but is entwined with inequalities in power at local, as well as national and global scales. The use of boxed examples, tables and illustrations helps students understand complex theoretical ideas and also demonstrates how development theories are put into practice in the real world. Each chapter ends with a summary section, discussion topics, suggestions for further reading and website resources.

Katie Willis is Professor of Human Geography and Director of the Centre for Developing Areas Research at Royal Holloway, University of London.

Routledge Perspectives on Development

Series Editor: Professor Tony Binns, *University of Otago*

The *Perspectives on Development* series provides an invaluable, up to date and refreshing approach to key development issues for academics and students working in the field of development, in disciplines such as anthropology, economics, geography, international relations, politics and sociology. The series will also be of particular interest to those working in interdisciplinary fields, such as area studies (African, Asian and Latin American Studies), development studies, rural and urban studies, travel and tourism.

If you would like to submit a book proposal for the series, please contact Tony Binns on j.a.binns@geography.otago.ac.nz

Published:

Third World Cities, 2nd edition
David W. Drakakis-Smith

Rural-Urban Interactions in the Developing World
Kenneth Lynch

Environmental Management and Development
Chris Barrow

Tourism and Development
Richard Sharpley and David J. Telfer

Southeast Asian Development
Andrew McGregor

Population and Development
W.T.S. Gould

Postcolonialism and Development
Cheryl McEwan

Conflict and Development
Andrew Williams and Roger MacGinty

Disaster and Development
Andrew Collins

Non-Governmental Organisations and Development
David Lewis and Nazneen Kanji

Cities and Development
Jo Beall

Gender and Development, 2nd edition
Janet Henshall Momsen

Economics and Development Studies
Michael Tribe, Frederick Nixson and Andrew Sumner

Water Resources and Development
Clive Agnew and Philip Woodhouse

Theories and Practices of Development, 2nd edition
Katie Willis

Forthcoming:

Global Finance and Development
David Hudson

Africa: Diversity and Development
Tony Binns, Alan Dixon, and Etienne Nel

Politics and Development
Heather Marquette and Tom Hewitt

Food and Development
E.M. Young

Natural Resource Extraction
Roy Maconachie and Dr Gavin M. Hilson

Children, Youth and Development, 2nd edition
Nicola Ansell

Climate Change and Development
Thomas Tanner and Leo Horn-Phathanothai

Religion and Development
Emma Tomalin

Development Organizations
Rebecca Shaaf

An Introduction to Sustainable Development, 4th edition
Jennifer Elliott

Theories and Practices of Development

Second edition

Katie Willis

Routledge
Taylor & Francis Group

LONDON AND NEW YORK

First published 2005
by Routledge
2 Park Square, Milton Park, Abingdon, Oxon, OX14 4RN

Simultaneously published in the USA and Canada
by Routledge
711 Third Avenue, New York, NY 10017 (8th Floor)

Second edition 2011

Routledge is an imprint of the Taylor & Francis Group, an informa business

Typeset in Times New Roman and Franklin Gothic by
Florence Production Ltd, Stoodleigh, Devon

British Library Cataloguing in Publication Data
A catalogue record for this book is available from the British Library

Library of Congress Cataloguing in Publication Data
Willis, Katie, 1968–.
Theories and practices of development/Katie Willis. – [2nd ed.]
p. cm.
Rev. ed. of: Theories and practices of development. 2005.
Includes bibliographical references and index.
1. Economic development. 2. Economic development – Social aspects.
3. Economic development – Political aspects. I. Title.
HD75.W55 2011
338.9001 – dc22   2010035279

ISBN: 978–0–415–59070–9 (hbk)
ISBN: 978–0–415–59071–6 (pbk)
ISBN: 978–0–203–84418–2 (ebk)

Contents

Plates

Figures

Tables

Boxes

Acknowledgements

Since *Theories and Practices of Development* came out in 2005, I have been thrilled to receive emails from students and colleagues all over the world. While not everyone agreed with my approach and analysis, it has been a real pleasure to find out how people have used the book. It was also very helpful to receive referees' comments on the proposed revised edition. In revising the book, I have incorporated some of the suggestions from both email correspondents and referees, but I am afraid I have not been able to address all of them.

I would like to express sincere thanks to students and staff in the Geography Department at Royal Holloway, University of London. Students on the Year 1 'Geographies of Development' course have been exposed to some of the revised sections of the book and their responses have been very helpful. Students on the MSc in Practising Sustainable Development and PhD students in the Centre for Developing Areas Research (CEDAR) have also provided insights or have suggested readings, which appear at various points. Finally, many thanks to my CEDAR colleagues for their good humour and support, as well as thought-provoking discussions about this thing called 'development'.

Andrew Mould at Routledge has been hinting about a second edition of *Theories and Practices of Development* for a couple of years. I am very grateful that he invited me to revise the book and for his encouragement during the process. Faye Leerink at Routledge has also provided excellent support. Jenny Kynaston in the Graphics Unit at Royal Holloway produced most of the figures, for which I would like to express my great thanks, particularly as she was dealing with the challenges of designing and producing 'Peru banners' at the same time. I would also like to thank Kelly Carmichael for the images used in Plates 4.2, 5.4, 5.6 and 7.5 and Still Pictures for their permission to reprint Plates 3.4, 4.4, 6.3, 7.2 and 7.3. The base map data for the world maps comes from Maps in Minutes.

Finally, I would like to take this opportunity to thank a number of people, both for their suggestions for the book and their friendship and companionship on travels in the UK and further afield: Kelly Carmichael, Vandana Desai, Dorothea Kleine, Emma Mawdsley, Claire Mercer, Paula Meth, Jay Mistry, Nicola Shelton and Glyn Williams.

Katie Willis
London, August 2010

Acronyms

ACAP	Annapurna Conservation Area Project
ACP	Africa, Caribbean and Pacific
AFP	Alliance for Progress
AGRA	Alliance for a Green Revolution in Africa
AIDS	Acquired Immune Deficiency Syndrome
ALBA	Bolivarian Alliance for the Peoples of Our America (acronym in Spanish)
AMISOM	African Union Mission in Somalia
APEC	Asia-Pacific Economic Cooperation
ASEAN	Association of Southeast Asian Countries
BNA	Basic Needs Approach
BPO	Business Process Outsourcing
BRIC	Brazil, Russia, India and China
CACM	Central American Common Market
CAMC	Conservation Area Management Committee
CARICOM	Caribbean Community
CARIFTA	Caribbean Free Trade Association
CDM	Clean Development Mechanism
CEDAR	Centre for Developing Areas Research
CEPAL	UN Economic Commission for Latin America (acronym in Spanish)
CET	Common Extended Tariff
CFC	Chlorofluorocarbon
CIS	Commonwealth of Independent States
CMEA	Council for Mutual Economic Assistance (Comecon)
CRC	Convention on the Rights of the Child

CRPD	Convention on the Rights of Persons with Disabilities
DAC	Development Assistance Committee
DDT	Dichloro-diphenyl-trichloroethane
DFID	Department for International Development
ECLA	UN Economic Commission for Latin America (acronym in English)
EPA	Economic Partnership Agreement
EU	European Union
EZLN	Zapatista National Liberation Army (acronym in Spanish)
FAO	Food and Agriculture Organization
FBDO	Faith-based Development Organization
FBO	Faith-based Organization
FDI	Foreign Direct Investment
FLO	Fairtrade Labelling Organizations International
G20	Group of 20 (19 economically dominant countries and the EU)
GATT	General Agreement on Tariffs and Trade
GB	Grameen Bank
GDI	Gender-related Development Index
GDP	Gross Domestic Product
GEM	Gender Empowerment Measure
GM	Genetically Modified
GNH	Gross National Happiness
GNI	Gross National Income
GNP	Gross National Product
HDI	Human Development Index
HIPC	Heavily Indebted Poor Countries initiative
HIV	Human Immuno-deficiency Virus
HPI	Human Poverty Index/ Happy Planet Index
HYV	High-yielding Variety
IBRD	International Bank for Reconstruction and Development
ICSID	International Centre for the Settlement of Investment Disputes
ICT	Information and Communication Technology
IDA	International Development Association
IFAS	International Fund for the Aral Sea
IFC	International Finance Corporation
IFI	International Financial Institution
ILO	International Labour Organization
IMF	International Monetary Fund

INGO	International Non-Governmental Organization
ISI	Import-Substitution Industrialization
ITDG	Intermediate Technology Development Group
LEDC	Less Economically Developed Country
LECZ	Low Elevation Coastal Zone
MAS	Movement Towards Socialism (acronym in Spanish)
MDG	Millennium Development Goal
MEDC	More Economically Developed Country
MERCOSUR	Southern Cone Common Market
MIGA	Multilateral Investment Guarantee Agency
MIPAA	Madrid International Plan of Action on Ageing
MNC	Multinational Corporation
MPI	Multidimensional Poverty Index
MST	Rural Landless Workers Movement (acronym in Portuguese)
NAFTA	North American Free Trade Agreement or Area
NAM	Non-Aligned Movement
NEP	New Economic Policy
NEPAD	New Partnership for Africa's Development
NGO	Non-Governmental Organization
NIC	Newly-Industrializing Country
NIDL	New International Division of Labour
NIE	Newly-Industrializing Economy
NLD	National League for Democracy
NPA	New Policy Agenda
NTNC	National Trust for Nature Conservation
OAU	Organization for African Unity
ODA	Official Development Assistance
OECD	Organisation for Economic Co-operation and Development
OEEC	Organisation for European Economic Co-operation
OPEC	Organization of Petroleum Exporting Countries
PES	Payment for Ecosystem Services/ Payment for Environmental Services
PRA	Participatory Rural Appraisal
PRI	Party of the Institutionalized Revolution (acronym in Spanish)
PRODEPINE	Project for the Development of Indigenous and Black People in Ecuador (acronym in Spanish)
PRS	Poverty Reduction Strategy
PRSP	Poverty Reduction Strategy Paper
PUA	Participatory Urban Appraisal

RAAN	North Atlantic Autonomous Region (acronym in Spanish)
SACU	Southern African Customs Union
SAP	Structural Adjustment Programme
SDR	Special Drawing Right
SEZ	Special Economic Zone
TNC	Transnational Corporation
UN	United Nations
UNDP	United Nations Development Programme
UNEP	United Nations Environment Programme
UNICEF	United Nations Children's Fund
USA	United States of America
USSR	Union of Soviet Socialist Republics
USAID	United States Agency for International Development
USSR	Union of Soviet Socialist Republics
USTR	United States Trade Representative
VHC	Viviendas del Hogar de Cristo
VSO	Voluntary Service Overseas
WCED	World Commission on Environment and Development
WTO	World Trade Organization

1 Introduction: what do we mean by development?

- Definitions of development
- Measuring development
- Colonialism
- Development actors
- Postcolonialism, postmodernism, post-development

In September 2000 United Nations members adopted the Millennium Declaration, out of which came the 'Millennium Development Goals' (MDGs) (see Box 1.1). Since then, these goals have been widely used by multilateral agencies, governments and non-governmental organizations (NGOs), in framing development policies in order to achieve the associated targets by 2015. Such clearly stated goals suggest that defining 'development' is easy and that what is important is the end point that a society gets to, not how those goals are achieved.

Box 1.1

Millennium Development Goals

While these goals were adopted by the UN in 2000, they were the outcome of international conferences throughout the 1990s. There are eight goals, but for each goal there are a number of targets and indicators. The eight goals are:

1 eradicate extreme poverty and hunger;
2 achieve universal primary education;
3 promote gender equality and empower women;
4 reduce child mortality;

5 improve maternal health;
6 combat HIV/AIDS, malaria and other diseases;
7 ensure environmental sustainability;
8 develop a global partnership for development.

The targets are much more specific and include:

1 between 1990 and 2015, halve the proportion of people whose income is less than US$1 a day;
2 reduce by two-thirds, between 1990 and 2015, the maternal mortality rate;
3 have, by 2015, begun to reduce the incidence of malaria and other major diseases;
4 halve, by 2015, the proportion of people without sustainable access to safe drinking water and basic sanitation.

Source: adapted from Development Goals (2010)

In this book we will be considering theories about development and how these theories inform policy formulation and practices to achieve development goals. However, before we embark on this journey, we need to consider what 'development' means. Despite the seemingly 'common sense' nature of the MDG 'development targets', this chapter will highlight the contested nature of the term 'development'. In particular, we will look at *how* 'development' has been defined, *who* has defined 'development' and at *what scale* 'development' has been examined.

Modernity

For many people, ideas of development are linked to concepts of modernity. 'Modernity' in its broadest sense means the condition of being modern, new or up-to-date, so 'the idea of "modernity" situates people in time' (Ogborn 2005: 339). Because of social, economic, political and cultural dynamism, what is 'modern' will change over time and also spatially. What is 'modern' in one place may be 'old-fashioned' elsewhere.

However, more specifically, 'modernity' has been used as a term to describe particular forms of economy and society based on the experiences of Western Europe and more recently the USA.

In economic terms, 'modernity' encompasses industrialization, urbanization and the increased use of technology within all sectors of the economy. This application of technology and scientific principles is also reflected within social and cultural spheres. What has been termed the 'Enlightenment' period in Western Europe in the late seventeenth and eighteenth centuries involved the growing importance of rational and scientific approaches to understanding the world and progress (Sheppard *et al.* 2009: 54–6). This was contrasted with previous understandings that were often rooted in religious explanations (Power 2003: 72–6). Approaches to medicine, the legal and political systems and economic development were all affected by this shift in perspective.

The spatial and temporal context of these ideas about modernity is important in this understanding of what 'modern' was, but as we shall see throughout the book, these ideas were taken out of their context and spread throughout the world (Larrain 2004). For some, this diffusion of modernity is interpreted as 'development' and 'progress', while for others it is associated with the eradication of cultural practices, the destruction of natural environments and a decline in the quality of life. All these themes, and others, will be considered in the following chapters.

Development as an economic process

People defining development as 'modernity', look at development largely in economic terms. This conception of development underpins much of the work of international organizations such as the World Bank, and also many national governments in both the Global North and Global South. The World Bank, for example, uses Gross National Income per capita (GNI p.c.) to divide the countries of the world into development categories. Low-income countries are defined as those with a GNI p.c. figure in 2008 of US$945 or less, lower-middle-income countries have US$946–3,855, upper-middle-income countries US$3,856–11,905 and high-income countries are those with GNI p.c. of US$11,906 or more (World Bank 2010e: 377) (Figure 1.1). GNI is a purely economically-based measure (Box 1.2). Because countries vary so greatly in population, the total GNI figure is divided by the number of people in the country, giving a per capita (p.c.) figure to indicate economic wealth.

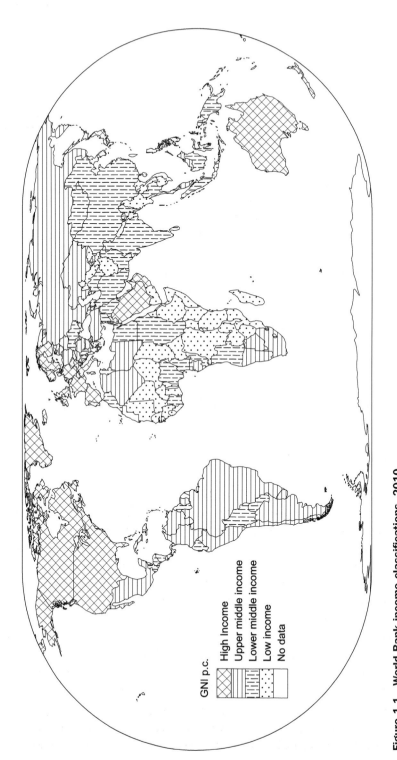

Figure 1.1 World Bank income classifications, 2010.

Source: based on data from World Bank (2010e: 377)
Map data © Maps in Minutes™ (1996)

GNI p.c.

High Income
Upper middle income
Lower middle income
Low income
No data

Box 1.2

Calculations of GDP, GNP and GNI

Gross Domestic Product (GDP)   This measures the value of all goods and services produced within a particular country. It does not matter whether the individuals or companies profiting from this production are national or foreign.

Gross National Product (GNP)   This measures the value of all goods and services claimed by residents of a particular country regardless of where the production took place. It is, therefore, GDP plus the income accruing from abroad (such as repatriation of profits) minus the income claimed by people overseas.

Gross National Income (GNI)   This is an alternative name for GNP. The World Bank now refers to GNI rather than GNP in its annual *World Development Report*.

The use of a wealth measure to represent development is regarded as appropriate because it is assumed that with greater wealth come other benefits such as improved health, education and quality of life.

Human development

The GNI p.c. or GNP p.c. indicator is still widely used, but this has increasingly been in conjunction with other broader indicators of 'development' which have highlighted non-economic dimensions of the concept. The most frequently used of these is the Human Development Index (HDI) which was devised by the United Nations Development Programme (UNDP) in the late 1980s. While the measure still has an economic aspect, there are other indicators of development relating to well-being (Box 1.3). Since 1990, the UNDP has published the *Human Development Report* every year. The HDI is used to divide the world's countries into those with very high, high, middle and low human development (Figure 1.2).

If you compare Figures 1.1 and 1.2 you can see that there are great similarities in the patterns. The countries of Western Europe, the USA and Canada, Japan, Australia and New Zealand all rank

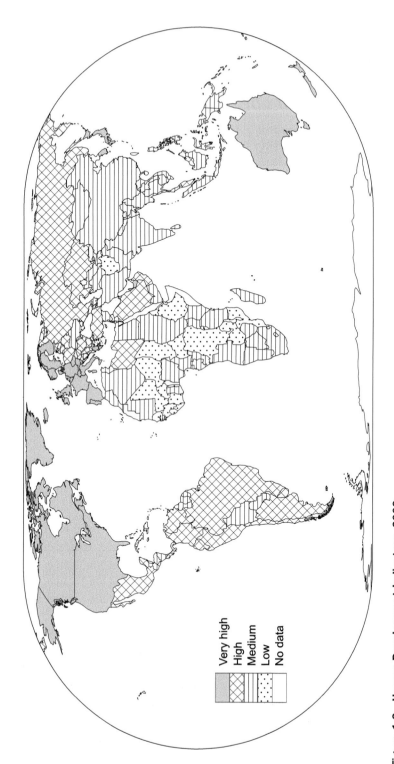

Figure 1.2 Human Development Indicators, 2009.

Source: based on data from UNDP (2009: 143–6)
Map data © Maps in Minutes™ (1996)

Very high
High
Medium
Low
No data

Box 1.3

Human Development Index

In the late 1980s increased awareness that the commonly-used economic measures of development were far too limited led the United Nations Development Programme (UNDP) to devise the Human Development Index (HDI). This measure incorporates three dimensions of development in relation to human well-being: a long and healthy life, education and knowledge, and a decent standard of living. The UNDP selected four quantitative indicators to measure these dimensions.

Calculation of the HDI

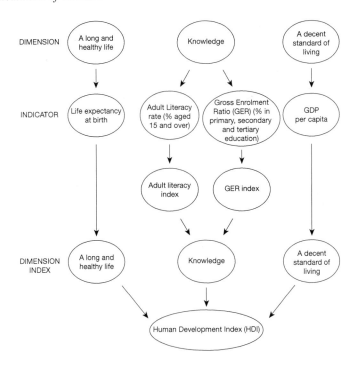

The indicators then have to be converted to an index from 0 to 1 to allow for equal weighting between each of the three dimensions. Once an index value has been calculated for each dimension, they are averaged and the final figure is the HDI. The higher the value the higher the level of human development.

Source: adapted from UNDP (2009: 206)

highly on both GNI p.c. and HDI figures. Similarly, most Southern African countries are classified as upper-middle-income countries with medium human development. Low GNI p.c. and low levels of human development at a national scale coincide in much of Central and West Africa. If GNI p.c. seems to present us with the same results does this mean that there is no real reason to use the more complex HDI measure? There are a number of reasons why this may not be appropriate. First, there is not complete overlap. For example, while Vietnam, Laos and Cambodia are categorized as 'low-income countries', their HDI scores put them in the category of 'medium human development'. In addition to the lack of complete overlap, by using the HDI you are asserting that 'development' is more than just economic progress measured at a national scale.

For some, however, these discussions of appropriate measures of national status are unimportant, because such measures do not consider inequalities in either spatial terms (see below) or in social terms. They also do not encompass how the vast majority of the people define development or how they would like their lives to change (if at all) (Friedmann 1992b).

The importance of scale

Development can be considered at a number of spatial scales. These go from the individual, to the local community, the regional, the national and the global (among others). How development is defined may differ by scale and, in addition, the approaches to development may be similarly scale dependent as we shall see in the next section.

Inequalities can be revealed at particular spatial scales. For example, if we consider national level development figures we get no idea of whether there are differences between regions within the country. As we shall see throughout the book, spatial inequalities are a key factor in any discussion of development. Some forms of development may lead to increasing inequalities between places, while other development approaches may explicitly attempt to reduce spatial inequalities.

At the sub-national scale, it is also important to recognize distributional issues. The Gini coefficient is a measure of inequality (see Box 1.4). At a national level, while income per capita levels and HDI may be 'satisfactory' according to international norms, it is important to recognize that not everyone in the country will have access to that level of income or standard of living (see Table 1.1). As this table shows, these issues of inequality are as important in the Global North as in the Global South – high levels of economic development do not necessarily mean great equality (see Box 1.5). In addition, experiences of marginalization, poverty and disadvantage are not restricted to certain parts of the world (Jones 2000).

A key sub-national pattern of spatial inequality is between rural and urban areas. If we consider indicators of economic and social

---

Box 1.4

Gini coefficient and Gini index

Both of these are measures of inequality and are named after the Italian statistician who formulated the coefficient in 1912. They measure either income inequality or inequalities in consumption between individuals, households or groups.

Gini coefficient   This measure varies from 0, which means perfect equality, to 1 which represents perfect inequality. Thus, the nearer the coefficient is to 0 the more equal the income distribution. Countries with a Gini coefficient of between 0.50 and 0.70 could be described as having highly unequal income distributions, while those with Gini coefficients of 0.20 to 0.35 have relatively equitable distributions.

Gini index   This measure, used by the UNDP, ranges from 0 to 100. A figure of 0 means perfect equality and 100 means perfect inequality.

Source: adapted from Todaro (2000); UNDP (2009)

Box 1.5

Inequality in the USA

With a GNP p.c. figure of US$47,580 in 2008, the USA is among the richest nations in the world. However, these average national figures hide massive inequalities in income and very different life experiences. With a Gini index of 40.8, it is clear that not all Americans have an equal share of the nation's riches. According to the US Census Bureau, in 1973 the top 20 per cent of earners in the US had 44 per cent of the total income. By 2000 this had increased to 50 per cent. Figures for all wealth, not just income, show a similar pattern of inequality, with the wealthiest 1 per cent of households controlling 38 per cent of the national wealth, while the bottom 80 per cent of households only controlled 17 per cent.

This economic inequality is also apparent in social indicators. Amartya Sen in his book *Development as Freedom* (1999), argues that comparing some groups within the US to societies in the Global South demonstrates that Americans can be in a worse position than their counterparts in poorer countries. While African-Americans in the USA earn far more than people born in China or Kerala (SW India), they have a lower chance of reaching advanced ages. Sen also uses the results of medical research by McCord and Freeman (1990) to state 'Bangladeshi men have a better chance of living to ages beyond forty years than African-American men from the Harlem district of the prosperous city of New York' (1999: 23).

Sources: adapted from *The Economist* (2003); Sen (1999); UNDP (2009); World Bank (2010e)

Table 1.1 *Measures of income inequality*

	HDI ranking 2009	Richest 10% to poorest 10%	Gini index
Australia	2	12.5	35.2
Japan	10	4.5	24.9
United States	13	15.9	40.8
Poland	41	9.0	34.9
Brazil	75	40.6	55.0
Turkey	79	43.2	17.5
China	92	13.2	41.5
India	134	8.6	36.8
Nigeria	158	16.3	42.9
Zambia	164	29.5	50.7
Ethiopia	171	6.3	29.8
Niger	182	43.9	15.3

Figures for the period 1992–2007

Source: adapted from UNDP (2009: 193–8)

well-being, there seems to be a clear trend of rural–urban inequality with rural populations generally being worse off than their urban counterparts (Table 1.2). However, such distinctions must be treated with caution (Wratten 1995). First, poverty indicators are notoriously problematic. For example, in a rural area,

Table 1.2 *Rural–urban differences in access to water and sanitation services*

	Urban population as % of total (2005)	Population with access to safe drinking water (%) (2004)		Population with access to improved sanitation services (%) (2004)	
		Urban	Rural	Urban	Rural
Mexico	76.0	100	87	91	41
Cuba	75.5	95	78	99	95
Botswana	57.4	100	90	57	25
Kazakhstan	57.3	97	73	87	52
Syria	50.6	98	87	99	81
Turkmenistan	46.3	93	54	77	50
China	40.4	95	78	99	95
India	28.7	95	83	59	22
Bangladesh	25.1	82	72	51	35
Cambodia	19.7	64	35	53	8
Rwanda	19.3	92	69	56	38

Source: adapted from World Bank (2009: 335–7)

monetary income may be lower than in the towns and cities, but the cost of living is lower and the availability of food from subsistence farming may help save on food costs. Second, the distinctions between rural and urban areas are never as distinct as statistics may imply. In most parts of the world, the linkages between rural and urban areas are multiple, with significant seasonal migration flows between the countryside and the city (Frayne 2010; Lynch 2005; Tacoli 2006). As cities have grown, the role of the peri-urban area has also become more important for food production and employment opportunities (McGregor *et al.* 2006). Finally, it must be remembered, that in some regions of the Global South, particularly Latin America and the Caribbean, the population is predominantly urban (Figure 1.3). Thus, while poverty levels may be higher in rural areas, poverty is increasingly an urban phenomenon because the majority of the population is urban (UN-Habitat 2010).

As will be discussed in much more detail in Chapter 5, inequalities are not just experienced in spatial terms, social inequalities are also very important. Throughout the world women as a group have tended to be excluded from many of the benefits which development of certain forms brings (Momsen 2010). Particular ethnic groups in regional or national contexts may also be deprived of opportunities,

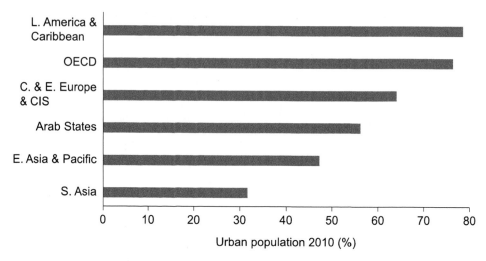

Figure 1.3 Urbanization levels by region, 2010.
Source: based on data from UNDP (2009: 194 and Table L Errata www.undp.org)

or may be denied decision-making power in the framing of development projects. This can lead to destructions of cultural practices and institutions, as well as a decline in self-respect and self-esteem. How to deal with social diversity is a key theme in development thought; not only in trying to implement development practice, but in actually defining what is meant by 'development'. Social diversity is dealt with throughout the book, but particularly in Chapter 5.

Measuring 'development'

It is not only defining 'development' which is contested, the way that development, regardless of definition is measured is also problematic. Of course, this assumes that 'development' is something which needs to be measured or assessed. For various actors in development (see pp. 26–7) measuring development could be important. For example, policy-makers may want to find out what the social development position (as defined by the policy-makers) of a population is in order to inform policy formulation. Governments or international agencies may want to assess the impact of a particular development initiative and therefore want to have measurements from both before and after the project. Finally, campaigning organizations seeking to improve living conditions for marginalized groups, may want information about the nature of marginalization.

Because of the different conceptions of development and the range of scales at which it can be considered, measuring 'development' requires proxies (Morse 2004). For example, in the case of the World Bank focus on economic development, the indicator used is GNI per capita. This has now been widely adopted as an appropriate measure of economic development, but other indicators or proxies could be used, such as the contribution of non-agricultural activities to GDP. In the case of the HDI, the UNDP decided that its understanding of human development included three main features: health, education and economic status. To measure each of these the UNDP needed to choose indicators (Box 1.3). This choice of indicators is certainly not straightforward. For example, Hicks and Streeten (1979) discuss the issues around trying to find proxy measures for 'basic needs' (see Chapter 4). While there may be agreement on what 'basic needs' are, including adequate shelter, food, clothing and employment, it is much more difficult to work out how these elements are to be measured.

Another problem with measuring development is comparability. This can be over time, or between different countries. Collecting large amounts of information, for example through national censuses, requires significant resources in terms of trained personnel and technology for analysing the results. These are clearly not equally available to all national governments (Bulmer and Warwick 1993). In addition, data collection can be disrupted by political unrest or war, and some communities or groups may be excluded from surveys and other studies because they are socially, economically or geographically marginal (Chambers 1997).

Finally, development measures are nearly always quantitative, i.e. they can be expressed in numerical form. This focus is understandable given the need to make comparisons across time and space, and also to deal with large amounts of information. However, by focusing on quantitative measurement, the subjective qualitative dimensions of development are excluded. This means excluding the feelings, experiences and opinions of individuals and groups. This approach also tends to reinforce outsiders' ideas about 'development', rather than what local people think 'development' is, or should be (Chapter 4).

A good example of this debate is the definition of 'poverty' (McIlwaine 2002; White 2008). The Millennium Development Goals have poverty reduction at their core. The definition of poverty used in these targets is an economic one and the measurement used is a poverty line. The original MDG target used US$1 per day as the international poverty line, but in 2008 this was revised to US$1.25 per day to reflect cost of living increases (World Bank 2008). However, this economic view of poverty is very limited and assumes a clear relationship between income poverty and other measures of disadvantage. Because of this, the UNDP devised the Human Poverty Index (HPI), which has been used since 1997. There are two slightly different measures; HPI-2 for 31 Organisation for Economic Co-operation and Development (OECD) countries (mainly Northern countries) and HPI-1 for 135 developing countries and areas, but both encompass indicators of health, education and standard of living (Table 1.3). These measures of poverty tend to be applied at a national scale.

A more recent attempt to measure poverty is the Multidimensional Poverty Index (MPI). This identifies health, education and living standards as key aspects and uses ten indicators to measure household poverty. These indicators include nutritional level, access to sanitation services and school enrolment. However, an additional feature of this measure is an assessment of the intensity of poverty, taking into account how many of the poverty indicators are found in a particular household. Because of the household level data, the MPI

Table 1.3 _Human Poverty Index_

Dimension	Measure
HPI-1 (for developing countries)	
Long and healthy life	Probability at birth of not surviving to age 40
Knowledge	Adult (aged 15 and above) illiteracy rate
Decent standard of living	% population without access to treated water supplies
	% children under five who are underweight
HPI-2 (for OECD countries)	
Long and healthy life	Probability at birth of not surviving to age 60
Knowledge	% adults (aged 16–65) lacking functional literacy skills
Decent standard of living	% people living below half the median disposable household income
Social exclusion	Rate of long-term (over 12 months) unemployment

Source: adapted from UNDP (2009: 206)

can be used to assess differences within countries and also between different social groups (Alkire and Santos 2010). The MPI has been used by the UNDP in the *Human Development Report 2010* (UNDP 2010a).

Despite the growing complexity of poverty measures, they still exclude any qualitative examination of experiences of poverty. Cathy McIlwaine (2002: 82) uses quotations to exemplify how poverty can be experienced and understood in different ways:

> 'For me, being poor is having to wear trousers that are too big for me.' (José, 8 years old, Guatemala City)
>
> 'Poverty makes my children get sick and they get worse because we're too poor to buy medicines.' (Antonia, 30 years old, Guatemala)
>
> 'It's poverty that makes me drink until I fall over, and drinking until I fall over makes me poor.' (Eduardo, 35 years old, Guatemala)

The qualitative examination of poverty puts the experiences of the people directly affected at the heart of the study. For some approaches to development this people-centred approach is key (see Chapter 4) and represents a move away from national-level considerations. Although the World Bank usually uses quantitative measures of development, in preparation for the 2000/2001 *World Development Report* which was on 'Attacking Poverty', it commissioned a large study entitled 'Voices of Poor' which attempted to examine the experiences of poverty throughout the world (Parnwell 2003). While the information gathered in this study was incorporated into the 2000/2001 *World Development Report*, there seems to have been a retreat back to quantitative measures since then (Williams and McIlwaine 2003). This discussion of poverty measurements shows how even the most 'basic' of 'development' measures is difficult to assess.

Terminology

The UNDP categorization of countries as having 'very high', 'high', 'medium' or 'low' human development based on HDI and the World Bank use of GNI per capita to place countries into one of four classes, are two examples of how the world can be divided up according to levels of 'development'. There are, however, many other forms of classification and a range of terms to describe groups of countries. Rather than merely being a debate about terminology which has no bearing on real-life issues, it is important to realize that the way that different parts of the world are described can tell us a great deal about who has the power to decide what should be valued and what denigrated. There has been growing awareness of how visual and textual representations of peoples and places both reflect prevailing power relations and reinforce certain ways of perceiving the world (Williams *et al.* 2009: Chapter 2). Postcolonial and post-development approaches (discussed later in this chapter) are particularly engaged with examining how certain forms of knowledge are validated while others are ignored, and the real-life effects of these processes.

In this book I will generally use the terms 'Global North' or 'North' to describe the countries of Europe, Japan, Australia, New Zealand, USA and Canada, and the 'Global South' or 'South' to describe the remaining countries of Africa, Asia, Latin America, the Caribbean and the Pacific. While there are clearly problems with using these terms, not least the fact that not all 'Northern' countries are north of the equator and not all 'Southern' ones south of the equator, I prefer to use these terms rather than other common distinctions discussed below. In addition, the North/South distinction was used by the Brandt Commission in its report on the nature of global interdependence (Brandt Commission 1980). The Commission, also known as the Independent Commission on International Development Issues, was set up in 1977 to consider issues of global inequality and poverty. It was chaired by the ex-Chancellor of West Germany, Willy Brandt.

The term 'Third World' has often been used to refer to the nations of Africa, Asia, Latin America and the Caribbean. It was originally used to describe those countries which were part of the Non-Aligned Movement (NAM), i.e. they did not officially support either the capitalist USA or the communist USSR during the cold war, instead preferring a 'third way'. Under this interpretation the 'First World' consisted of the industrialized capitalist nations of Western Europe, the USA, Canada, Japan, New Zealand and Australia, while the 'Second World' was the communist bloc of the USSR and Eastern Europe. However, despite not originally having a sense of hierarchy, the idea of 'First', 'Second' and 'Third' was often interpreted as meaning the countries in the 'Third World' were in third place. The collapse of the 'Second World' in the late 1980s/early 1990s, with the transition from state-socialism, has meant that the basis for the distinction has been removed (see Chapter 3) (Friedmann 1992b).

Another popular form of constructing categories is the 'developed'/'developing' binary. This was felt to be better than distinguishing between 'developed' and 'undeveloped', as the latter phase implied being unable to escape from the condition of lack of development, rather than the more positive sentiment which 'developing' suggests. However, for some theorists (such as Frank 1967) the concept of being 'fixed' or unable to escape from a position of disadvantage because of global inequalities means that the term 'undeveloped' or 'underdeveloped' is crucial (see Chapter 3). In both 'developed'/'developing' and 'developed'/ 'undeveloped' the first term remains the same and seems to imply that 'development' is an end point, i.e. once you reach a certain standard of living or economic position then you are 'developed'. Again, such notions are important in some theorizing (such as modernization approaches discussed in Chapter 2). However, this fails to recognize the dynamism of all societies and the continued desire by populations for improvements (not necessarily in material goods). It also fails to consider the experiences of social exclusion that are found within supposedly 'developed' countries or regions (Jones 2000).

The terms 'More Economically Developed Countries' (MEDCs) and 'Less Economically Developed Countries' (LEDCs) have also gained in popularity. The explicit reference to *economic* development does not assume that development is automatically economic, or that economic development is necessarily associated with other forms of development. While this specificity is welcome, the emphasis on the economic, rather than other possible dimensions of development, could be regarded as implying that economic factors are the most important aspects of 'development'. As with all the categories used, where the boundary between groupings is placed is highly contentious, not least because of the rise of certain countries including the Gulf States and Brazil, Russia, India and China (collectively known as the BRICs).

Finally, some political activists working for greater global justice, refer to Africa, Asia, Latin America and the Caribbean as the 'Majority World' and the rest of the world as the 'Minority World'. The *New Internationalist* magazine uses this terminology, for example, to stress the fact that in population terms, the majority of the world's population (just over 80 per cent in 2008 according to World Bank 2010e: 379) lives in the nations of what I have termed the 'South'. This is an important point to make, as it stresses the Eurocentric assumptions which underlie many terms used.

The concept of 'Eurocentrism' will be important throughout the discussions of development theories. It refers to the assumption that European or Western ideas are the only ideas or approaches that are important. In some cases, this is because the theorist does not see that their approach is very context specific and that in fact there could be other interpretations, but in many other cases the Eurocentrism is based on ideas of Western/Northern superiority (Blaut 1993). Of course the concept of 'Eurocentrism' is also based on the assumption that the 'North' is homogenous. This is clearly not true given the range of nations making up the 'North', but also because of distinctions based on gender, ethnicity, class and many other social characteristics. Eurocentrism implies having power over knowledge, and because of this is regarded as reflecting existing class, gender and ethnic power relations such that the opinions of 'White', middle-class or elite men in the North are privileged.

It is not only terminology which can reveal biases and assumptions, maps are also important bearers of ideas because they are representations of the world. Because we all have different views of the world, how we choose to present our world in a map can reveal a great deal about our own particular biases (Wood 2010). A map projection is a way of portraying a three-dimensional globe on a flat piece of paper. Eurocentric maps, such as those drawn using the Mercator projection, place Europe at the centre of the map and represent the continents in the same shapes as they are in reality (Figure 1.4). However, because the Earth is a sphere this leads to the land masses nearer the poles appearing much larger relative to other continents nearer the equator. The Peters projection is an attempt to challenge this Eurocentric image. The Peters projection is an equal area projection, meaning that the land area represented on the map is correct in relation to other land areas. This means that Africa, Asia and Latin America are much more significant in the Peters projection, reflecting their importance in area terms in reality (Figure 1.5). Because of this, the Peters projection has often been used in development education schemes to try and counter

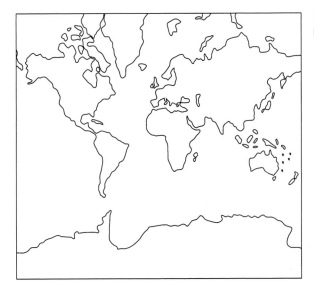

Figure 1.4 Mercator projection.

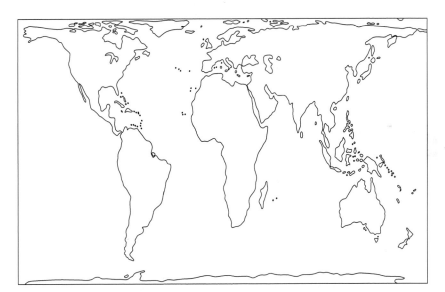

Figure 1.5 Peters projection.

Eurocentric bias (Vujakovic 1989). The Peters projection has, however, been criticized for making the continents appear long and thin, very unlike their shapes on the globe. In this book the world maps are drawn using the Eckert IV projection. This is an equal area projection which tries to minimize distortions to the shapes of the continents.

Colonialism

At the start of the twenty-first century there are very few colonies left in the world following widespread decolonization processes after the Second World War (however, see McEwan 2009: 19 for a list of 'overseas territories'). Despite this, any consideration of development theories and practices needs to include a discussion of the importance and nature of colonialism. Before elaborating on these reasons, a definition of 'colonialism' is needed. Bernstein (2000: 242) defines colonialism as 'the political control of peoples and territories by foreign states, whether accompanied by significant permanent settlement . . . or not'. This political control represents global power differentials and is associated with dominance in other spheres such as the economy and cultural practices.

Three main reasons for discussing colonialism in the context of development theory can be identified. First, from the middle of the sixteenth century onwards, European colonialism created more and more linkages between different parts of the world. As we shall see throughout the remainder of the book, interactions at a global scale and the bonds between different regions and countries are referred to in a range of ways in a number of development theories. While the linkages between different parts of the world cannot be solely attributed to the operation of colonialism, it was a key element in developing the basis for what we now call 'globalization' (see Chapter 7).

A second important reason for considering colonialism in a book on development theories is the nature of power relations embedded in colonial processes. The expansion of European political, economic and social control over other parts of the world represented the greater power held by these nations (see pp. 21–3). In some development theories, these power inequalities between North and South help explain differential development experiences, with colonialism bringing beneficial changes to Northern countries, at the expense of those in the South (see Chapter 3). It is argued that these inequalities also continue to limit the autonomy of Southern countries and peoples to determine their own futures through processes of what has been termed 'neo-colonialism'. This term is used to describe global relationships which reflect the dominance of the North over the South, despite legal independence. It is used, for example, in relation to the influence of transnational corporations (TNCs) over the economies of the South (see Chapter 7), or the ability of

**Plate 1.1 The Dutch Church,
Melaka, Malaysia.**

Credit: Katie Willis

Northern governments to intervene in Southern governments'
decision-making through the workings of multilateral organizations
such as the World Bank (see Chapter 2).

Finally, the colonial experience varied across the world, depending
on the colonial power, pre-existing social, economic and political
structures in the colony, and the timing of the colonial encounter
(Bernstein 2000). Whatever the experience, it is clear that
colonialism changed the social structures, political and economic
systems, and cultural norms in many places both North and South.
The legacy of these changes continued into independence.

While colonialism is usually considered to be a European-led
phenomenon, the dominance of some societies over others dates
from before European excursions into Asia, Africa, Latin America
and the Caribbean (Williams *et al.* 2009: Chapter 3). For example,
the Aztec and Inca empires in Latin America were able to dominate
other groups and territories and use them for resources. Similarly
the Mogul empire (1526–1761) in what is now north-west India
was built on the gathering of tribute and taxes from peasants (Bujra
2000). Throughout Africa there were significant empires, such as
the empire of the Kush in the Nile Valley and a number of Islamic
empires in West Africa (Stock 2004). The expansion of Western
European influence had, however, much more widespread and
long-standing effects.

Plate 1.2   Teotihuacán, Mexico.
Credit: Katie Willis

The first main period of European colonial expansion was led by the Spanish and Portuguese in Latin America and the Caribbean following Columbus' arrival in the Americas in 1492. In the eighteenth century, Spanish influence also extended northwards to what are now the southern states of the USA (Plate 1.3). During the sixteenth and seventeenth centuries, the Spanish and Portuguese used what they called the 'New World' as a source of raw materials, silver in particular. There was some settlement, but overall the colonial project of both these nations was focused on mercantile activities (trade).

In the latter part of the seventeenth century, the Dutch and British came to the fore. While they did have some activities in Latin America and the Caribbean, much of their activity was focused in North America and South and East Asia. While the importance of trade for these colonial endeavours was still high, in particular tobacco from North America and spices and silks from Asia, as manufacturing became more important in Britain, the provision of raw materials for these industries took on more significance. Imports of cotton from North America were transported to the burgeoning

Plate 1.3   Spanish mission church,
San Juan Bautista, California.

Credit: Katie Willis

textile factories of Northern England, and tobacco and sugar were also processed. The slave trade was key in the expansion of cotton, tobacco and sugar production as slaves were the mainstay of the plantation workforce. Within Sub-Saharan Africa, European enclaves were found along the western coast where slave trading took place. For example, the British had bases in Gambia, Sierra Leone, and the French in Senegal. Europeans (British, Dutch, French and Germans) settled in South Africa in 1652 in what is now Cape Town (Stock 2004). This period of colonialism also differed from the earlier Spanish and Portuguese phase because there was greater settlement by Europeans and the colonies became important markets for European manufactured goods (Bernstein 2000).

As industrial expansion took hold in Europe in the eighteenth and nineteenth centuries, colonies became increasingly important as sources of raw materials and markets (see Chapter 3 for a discussion of Marxist interpretations of colonialism at this time). Spain and

Portugal were losing their positions as key colonial powers and at the start of the nineteenth century wars of independence broke out in much of Latin America, leading to independence for many of the Latin American nations in the 1820s. Meanwhile, British and French colonies in South and East Asia continued to thrive. Within what became South Africa, the Cape was annexed by the British in 1795. This prompted what was known as the 'Great Trek' in the 1830s and 1840s when thousands of Boers (Dutch White settlers and descendents) moved north and established the Boer republics of Transvaal and Orange Free State (Stock 2004). European colonization of most of the African continent only really took place in the latter parts of the nineteenth century during what became known as the 'scramble for Africa'. At the Berlin Conference of 1884–5, the European powers divided up the continent, agreeing that if countries could demonstrate 'effective control', then they could legally claim that territory (Stock 2004). Britain and France were again the key players, but Belgium, Portugal and Germany also gained territories (Figure 1.6).

Following the Second World War, the pressure for decolonization in Africa, Asia and the Caribbean increased for a number of reasons. The war had caused major economic problems in Western Europe. The two main colonial powers, France and Britain, had to turn to the USA for assistance (see Chapter 2). In addition, the new global super-powers the USA and Soviet Union (USSR) both advocated decolonization, not least because it would provide new opportunities for the spread of their own influences. These factors external to the colonies were complemented by the increasing calls for independence from the populations of the colonies themselves. Changing economic processes and the growing power of multi-national corporations (MNCs) also helped. Direct political control was no longer necessary for goods to be traded between countries (Potter *et al.* 2008). The combination of these factors led to a gradual process of decolonization.

Despite the achievement of political independence, the autonomy of the newly-independent states was certainly not achieved. Economic linkages, in particular, continued to keep the ex-colonies in a subservient or dependent position (see Chapter 3). It can also be argued that this process of neo-colonialism also extends to the continued representation of 'Western' or 'Northern' ways of doing things as 'better'. This is a key concern of many development

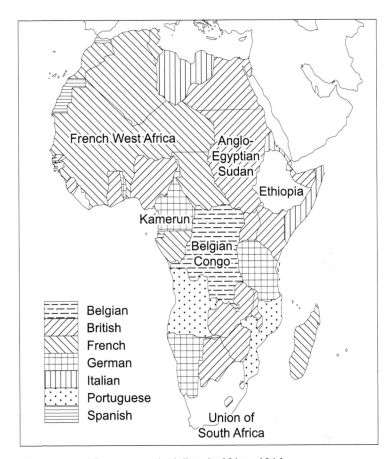

Figure 1.6 The extent of European colonialism in Africa, 1914.

Source: adapted from Simon (1994)

Map data © Maps in Minutes™ (1996)

theories outlined in the book and also post-development and postcolonial approaches, which are discussed at the end of this chapter.

The drawing together of different parts of the world through the political and economic processes associated with colonialism was just the start of the flows of ideas, commodities and people across the globe. This increasing interconnectedness is now referred to as 'globalization' (Chapter 7). However, just as during the colonial period certain countries and peoples were able to dominate others, so globalization reflects continuing power inequalities. Globalization is not experienced the same way by all the world's peoples. For example, certain parts of the world are more 'connected' to other regions than others. Sub-Saharan Africa, for example, is markedly less connected to other regions through trade, investment flows and telecommunications (Dicken 2007). This, some have argued (see Chapter 7), leaves the region unable to benefit from foreign investment and industrialization which are associated with economic globalization. The relative lack of connectedness in this sense, does not mean that the governments of Sub-Saharan Africa are necessarily able to adopt the development policies they feel are appropriate for their peoples. The importance of global institutions such as the World Bank (Chapter 2) and Northern governments, especially the USA, influences decisions at a national level. The colonial period may be over in formal political terms, but the inequalities in power and influence remain.

Actors in development

A key theme of this book is to understand how different definitions of, and approaches to, 'development' are linked to particular policy approaches. While the academic debates about 'development' are fascinating, it is particularly important to consider how these debates link to actual policies 'on the ground' which affect millions of people throughout the world.

The variety of approaches involves a range of actors, with shifting emphases being placed on these actors depending on the approach adopted. The degree of agency which these actors are perceived to have will also be affected by a particular interpretation of power distributions. Having agency implies that an individual or group is able to make decisions and do things based on their own choices (Garikipati and Olsen 2008). The other extreme, having no agency, means that there is no free will and individual behaviour is controlled by other actors.

There are a range of actors involved in development (Table 1.4). They vary from individuals to large-scale global organizations such as the United Nations. The scale differences are apparent, but it is important not to assume that there is an increase in influence as the scale increases. For example, individuals can be incredibly influential on a large scale because of their political or economic position, but individuals can also have very little influence even within their own households. The president of the United States of America and a woman farmer on the slopes of Mount Kilimanjaro in Tanzania are

Table 1.4 *Actors in development*

Actor	Activities
Individual	Depending on income, class, gender, ethnicity, age and other social variables can have a great deal of choice and influence, or be left with very little agency
Household	Group of people who live together and share expenses; not always members of the same family; can operate as a unit to ensure that all household members have their basic needs met
Community	Group of people with shared interests in some senses; usually based on shared residential location, e.g. a village or urban district, but can also refer to a community based on shared social identity
Government	Operates at a range of scales from local and municipal government to national government; important in setting economic framework; can be interventionist, or can play a regulatory role in development
Non-Governmental Organizations (NGOs)	Organizations which are neither run by the state nor profit-making companies; can help local communities set up projects to provide services, create income-generating opportunities, or improve social relations; can be very small-scale organizations, or very large global organizations such as Oxfam or Médecins Sans Frontières
Private companies	Representatives of the market; can be very small businesses or global corporations
Multilateral organizations	Can set global agenda for economic policies; promote global peace; important sources of aid and technical assistance. Examples: International Monetary Fund, United Nations, World Bank

both individuals, but their ability to influence events and their life choices are very different in scope.

Approaches to development

While one of the aims of this book is to highlight the complexities of the debates about 'development', as a starting point it is useful to have some basic framework within which to locate our discussions. Table 1.5 provides a chronology of 'development' approaches and understandings. The point of this table is not to suggest that theories of development have evolved in a unilinear way with no contestation

Table 1.5 *Main approaches to development, 1950s onwards*

Decade	Main development approaches
1950s	Modernization theories: all countries should follow the European model
	Structuralist theories: Southern countries needed to limit interaction with the global economy to allow for domestic economic growth
1960s	Modernization theories
	Dependency theories: Southern countries poor because of exploitation by Northern countries
1970s	Dependency theories
	Basic needs approaches: focus of government and aid policies should be on providing for the basic needs of the world's poorest people
	Neo-Malthusian theories: need to control economic growth, resource use and population growth to avoid economic and ecological disaster
	Women and development: recognition of the ways in which development has differential effects on women and men
1980s	Neoliberalism: focus on the market. Governments should retreat from direct involvement in economic activities
	Grassroots approaches: importance of considering local context and indigenous knowledge
	Sustainable development: need to balance needs of current generation against environmental and other concerns of future populations
	Gender and development: greater awareness of the ways in which gender is implicated in development
1990s	Neoliberalism
	Post-development: ideas about 'development' represent a form of colonialism and Eurocentrism. Should be challenged from the grassroots
	Sustainable development
	Culture and development: increased awareness of how different social and cultural groups affected by development processes
2000s	Neoliberalism: increased engagement with concepts of globalization
	Sustainable development
	Post-development
	Grassroots approaches
	Rights-based development

or conflict. Instead, as the following chapters will demonstrate, numerous ideas about 'development' can co-exist, although some theories will be adopted more widely, partly because they are advocated or supported by more powerful actors.

The table only covers development theories in the period after the Second World War. This is not because there were no ideas about social and economic development before then, but because in the 1940s and 1950s there was increasing international discussion about how 'development', particularly in the Global South, was to take place. International organizations were set up to try and achieve 'development' and a number of strategies were adopted. These specific interventions as part of an international development endeavour are what Gillian Hart (2001) terms '"big D" Development', in contrast to '"little d" development', which she sees as the general progress of capitalism. However, despite the focus of this book on post-Second World War theories and practices, as you will see, many of the ideas about development in the second half of the twentieth century and the start of the twenty-first had their roots in theorizing in the nineteenth century and earlier.

One feature of the chronological approach which should be highlighted is the concept of an 'impasse' in development theory (Schuurman 1993). In the 1980s, this idea of an impasse became increasingly common. In the 1960s and 1970s the contrasting approaches of modernization theories (see Chapter 2) and dependency theories (see Chapter 3) represented differing perspectives on development. However, the global economic problems of the 1980s and the awareness that in many senses existing 'development' theories had not been translated into practical success, led theorists to stop and think about what development was and how it could be achieved. While neoliberal thinking now dominates development policy-making (see Chapter 2), the post-1980s period has been associated with a recognition of much greater diversity within conceptions of development. This has included greater awareness of environmental concerns, gender equity and grassroots approaches. All these will be discussed in later chapters.

Postcolonialism and Postmodernism

Engaging with social diversity and also recognizing the importance of power relations in the construction and diffusion of development ideas, have been greatly associated with postcolonial and postmodernist approaches from the 1980s onwards. While the two approaches have similarities, they do not overlap completely. Postmodernism is difficult to define because it can be applied in a number of fields and in a variety of ways (Simon 1998). In the context of 'development' it has been particularly important in considering the ways in which previous understandings of 'development' assumed that the populations of the South were homogenous and that the European route to development was the only correct way.

The deconstruction of development categories is a key part of postmodern approaches to development. Rather than assuming that all 'peasants' are the same or all rural–urban migrants have the same experiences, postmodernism stresses diversity in social, spatial and temporal terms. For example, Chandra Talpade Mohanty (1991) focuses on the ways in which the term 'Third World Women' is used to describe all women living in the Global South. In particular, she highlights how this term is used to homogenize women's lives and is also used in a way that always implies victimhood; 'Each of these examples [in her chapter] illustrates the construction of "third world women" as a homogeneous "powerless" group often located as implicit *victims* of particular socioeconomic systems' (1991: 57, emphasis in the original). She argues that this approach not only denies the experiences of millions of women, but also reflects the power relations that frame understandings of the world, a key theme of postcolonial thought.

Postcolonial approaches seek to disrupt ways of thinking about the world based on Northern assumptions and also to recognize difference, but this is particularly within the context of places and peoples who have experienced colonialism from the perspective of being colonized. The term 'post-colonialism' is usually used to indicate a time period after colonialism, while 'postcolonialism' describes an approach to understanding social, economic, political and cultural processes (Loomba 1998). This includes both the material legacies of colonialism, such as urban structures and social hierarchies, as well as the how particular forms of knowledge are valued at the expense of others (Radcliffe 2005). For example, Frantz Fanon's book *Black Skin, White Masks*, originally published in French in 1952, highlights the effects of European colonialism on the mentalities of colonized Black populations. Postcolonialism therefore attempts to understand not only the observable legacies of colonialism, but also the ideas or discourses about 'development' that have been transferred as part of the colonial process (McEwan 2009).

Edward Said's book *Orientalism* (1991 [1978]) is an excellent example of postcolonialism. The book is subtitled *Western Conceptions of the Orient* and deals with how 'the West' has constructed the peoples of 'the East' as being 'backward' and 'uncivilized'. This has been used as a justification for political interventions and colonial projects. *Orientalism* shows how these ideas are constructed by particular groups of people at particular times, i.e. they reflect global power relations. In addition, Said also demonstrates how the construction of the 'East' as 'Other' and 'different' to the 'West' not only gives the 'East' a particular identity, but also reflects on the identity of the 'West' (Mercer *et al.* 2003).

Postmodern and postcolonial approaches to development have received some criticisms, in particular theorists are accused of 'playing academic games', rather than dealing with the day-to-day problems that millions of the world's poorest people face (Nederveen Pieterse 2000; Simon 1998; Sylvester 1999). These criticisms have also been levelled at the related 'post-development' ideas (see below). However, the importance of recognizing diversity in constructing development theories and practices is clearly of great importance, as is an awareness of the context in which theories are formulated. These themes will be developed throughout the rest of the book.

Post-development?

Alongside the debates about how 'development' can be achieved, since the 1990s, the concept of 'post-development' has come to the fore. One of the most well-known proponents of this approach is Arturo Escobar, who uses the case study of Colombia to discuss the development process. By 'development' he means the highly technocratic approach adopted by the World Bank, US government and other Northern institutions in the post-Second World War period (discussed in Chapter 2). His argument is that before 'outsiders' came into Colombia, there was no such thing as 'poverty' and therefore no need for 'development'. While most people had what would be defined as low life expectancies, many children lacked access to formal education and houses lacked water and electricity, these factors were not usually regarded as problems. Escobar argues that by imposing external norms and expectations on Colombian society and economy, the country was interpreted as 'lacking development'. This lack could only be addressed by adopting Northern forms of 'development'; hence numerous types of intervention in the form of aid and technical assistance (Chapter 2).

What Escobar and other post-developmentalists (see Rahnema with Bawtree 1997; Sachs 1992) argue, is that the development process as it has been experienced by Southern countries is based on Eurocentric assumptions. 'Development' has helped incorporate large areas of the globe into a Northern-dominated economic and political system which has destroyed indigenous cultures, threatened the sustainability of natural environments and has created feelings of inferiority among people of the South (Box 1.6). Post-development theorists stress the importance of the discourse of development. This refers to the way that 'development' is defined and discussed. Rather than being neutral, these theorists argue that understandings of 'development' reflect prevailing power relations and enable some ideas of 'development' to be presented as 'correct', while others are dismissed. As Cheryl McEwan (2009: 146) states, in the context of postcolonialism, 'Development discourse promotes and justifies very real interventions with real consequences.'

Summary

- Development is a highly-contested concept.
- Multilateral agencies often use economic measures such as GNI or GNP per capita to assess development.
- National-level measures hide important spatial and social inequalities.
- Despite widespread decolonization, it is important to consider the role of colonialism in understanding development today.
- Development as a process is not confined to Africa, Asia, Latin America and the Caribbean.
- Development can be understood as a Eurocentric idea which has been forced on the rest of the world.

Box 1.6

Nanda Shrestha's perspectives on development in Nepal

Nanda Shrestha is now Professor in the School of Business and Industry at the Florida A&M University, but he grew up in the 1940s and 1950s in Pokhara in central Nepal. His family survived by cultivating non-irrigated crops for subsistence and selling millet liquor. Hunger was common and their small house let the rain in. According to present-day assessments, his family and the wider community would certainly be classified as very poor and disadvantaged.

However, for Shrestha, the perception of their situation was very different:

> To my innocent mind, poverty looked natural, something that nobody could do anything about. I accepted poverty as a matter of fate . . . I had no idea that poverty was largely a social creation, not a bad karmic product. Despite all this, it never seemed threatening or dehumanizing. So, poor and hungry I certainly was. But underdeveloped? I never thought – nor did anybody else – that being poor meant being 'underdeveloped' and lacking human dignity. True, there is no comfort and glory in poverty, but the whole concept of development (or underdevelopment) was totally alien to me.
>
> (1995: 268)

In 1951, after a change of ruler in Nepal, western-funded development projects were introduced. The concept of development in Nepali is *bikas*. Shrestha describes how people were 'seduced' by this concept and saw everything that was associated with *bikas* as being good and of value, and everything else which was associated with existing ways of life as being inferior. This included forms of traditional medicine, manual labour, language and education. *Bikas* was regarded as desirable because it bought paved roads, school buildings and technology, even though hunger persisted and self-reliance and autonomy declined. Shrestha interprets this process as a form of colonialism, where European and American ideas and cultures are presented as being superior to indigenous ways of living.

Source: adapted from Shrestha (1995)

Discussion questions

1 Outline the major patterns of Human Development Index scores and suggest reasons for the differences between global regions.

2 In a postcolonial world, why is it important to consider colonialism in the context of development?

3 How do definitions of development vary according to scale?

4 What are the advantages and disadvantages of using quantitative measures of development?

5 What are the main features of a postcolonial approach to development?

Further reading

Esteva, G. 'Development' in W. Sachs (ed.) (1992) *The Development Dictionary: A Guide to Knowledge as Power*, London: Zed Books, pp. 6–25. An impassioned critique of 'development' as it has been defined in the twentieth century by policy-makers, particularly in the North. A useful introduction to the ideas of post-development.

Friedmann, J. (1992b) 'The end of the Third World', *Third World Planning Review* 14 (3): iii–vii. Clearly-written overview of the use of the term 'Third World' and why Friedmann believes it is no longer useful.

Jones, P.S. (2000) 'Why is it alright to do development "over there" but not "here"? Changing vocabularies and common strategies of inclusion across "First" and "Third" Worlds', *Area* 32 (2): 237–41. A short article which considers why 'development' is often only considered within the context of the Global South when there are problems of inequality and marginalization within Northern contexts as well.

McEwan, C. (2009) *Postcolonialism and Development*, Abingdon: Routledge. A clearly-written introduction to postcolonial theory and engagement with development theory and practice.

Rist, G. (2008) *The History of Development: From Western Origins to Global Faith*, 3rd edition, London: Zed Books. An excellent overview of how the concept of development has changed from the Enlightenment to the Millennium Development Goals.

Williams, G. *et al.* (2009) *Geographies of Developing Areas: The Global South in a Changing World*, Abingdon: Routledge. An accessible introduction to globalization and development. Chapter 2 provides a clear introduction to debates around representations of the Global South.

Useful websites

www.developmentgateway.org   Development Gateway. Links to a range of development information.

www.developmentgoals.org   World Bank Millennium Goals website. Details on what the goals are and what progress has been made.

www.eldis.org   Portal for development-related information run by the Institute of Development Studies, University of Sussex.

www.ophi.org.uk   Oxford Poverty and Human Development Initiative. Provides information about the concept of human development and also the Multidimensional Poverty Index (MPI).

www.un.org/millenniumgoals   United Nations Millennium Development Goals site.

www.worldbank.org/poverty   World Bank information and research on poverty reduction and equity.

2 Modernization, Keynesianism and neoliberalism

- Adam Smith and the free market
- Keynesianism
- Modernization theories
- Development aid
- Neoliberalism
- Structural adjustment policies
- Good governance
- Neo-Keynesianism and global economic crisis

This chapter deals with development theories that focus on the central role of the market in promoting economic progress. The role of other development actors, most notably the state, should not be ignored, but in most cases the state is viewed as an enabler for more effective market operations.

Classical theories

While the bulk of this book focuses on post-1945 development theories and policies, it is important to recognize that these ideas did not appear in an intellectual vacuum, but rather were rooted in the tradition of economic, political and sociological theorizing which developed in Europe from the eighteenth century onwards (Martinussen 1997: Chapter 2).

One of the key theorists to influence later ideas about economic development was Adam Smith. His book, *An Inquiry into the Nature and Causes of the Wealth of Nations* was published in 1776 and was a response to the mercantile (trade) focus of economic policy at that time in Western Europe. In the eighteenth century, it was trade which was the major force for economic growth; merchants, and particularly the large trading companies (such as the East India Company), had great power in relation to national governments. In order to safeguard their interests, merchants supported protectionist measures which allowed them to carry out their activities without what they saw as unnecessary competition. Protectionism included high import tariffs for goods produced outside a country. This made it cheaper for customers to buy domestically-produced goods.

Adam Smith viewed this form of regulation as detrimental to the economic growth of the country and greater wealth for all citizens, rather than just the merchant classes. He argued for greater attention to be paid to production, rather than trade, in economic development. In addition, he claimed that divisions of labour would help improve productivity and therefore economic growth and wealth creation. Division of labour describes the breaking up of the production process (for example making cloth) into a number of stages; rather than one person completing all stages, different people concentrate on one aspect of the process. They become greatly skilled at this and so more items can be produced in the same time.

The operation of the proposed system, he argued, would be regulated by the 'invisible hand of the market' rather than by the state (see Chapter 1 for a discussion of these actors). Smith believed that individuals would act in self-interest; thus if a product was too expensive then nobody would buy it and the seller would either reduce the prices or change to selling something else. Similarly, if wages were too low, then workers would move to other jobs. Despite writing before the turmoil and 'economic development' of the Industrial Revolution, Smith's work is still very influential today because of his theorizing about the role of the market in economic development. The market-centred approach to economic development has also been termed laissez-faire economics.

Another highly influential classical economist was David Ricardo, who lived in the late eighteenth to early nineteenth centuries. He was a great advocate of free trade and developed the theory of 'comparative advantage'. According to this theory, countries should concentrate on producing and then selling the goods that they had an advantage in producing because of their assets, such as land, mineral resource, labour, technical or scientific expertise. This meant a global division of labour. Ricardo argued that it made more sense for countries to specialize in this way, rather than trying to produce everything, because through specializing, production would be more efficient, there would be greater capacity for growth and scarce resources could be used more effectively (see Figure 2.1).

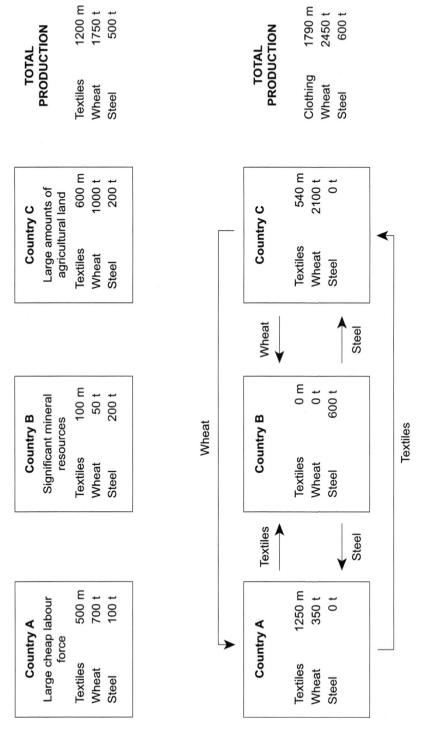

Figure 2.1 Comparative advantage.

Economic growth theory

Classical economists' belief in the market as a mechanism for maximizing efficient resource use and human well-being, was challenged in the early twentieth century by significant economic events, in particular the 1929 Wall Street Crash and the Great Depression of the 1930s in the USA. From this extreme failure of the free market to reach an equilibrium, economists began to develop new understandings of national economies. Foremost among these was the British economist John Maynard Keynes, who in 1936 published *The General Theory of Employment, Interest and Money*. Keynes' argument was that the free market was not necessarily the positive force that many, following Adam Smith, believed. Keynes argued that the key to growth was real investment, i.e. investment in new (rather than replacement) infrastructure projects. This investment, he claimed, would have a positive effect on job creation and the further generation of wealth, through the multiplier effect (see Figure 2.2). This effect could, however, also work in reverse, so that declining levels of real investment would lead to a downward spiral into economic crisis.

Unlike the classical economists, Keynes saw a key role for the government in promoting economic growth. Rather than letting the market operate alone, Keynes said that governments could intervene

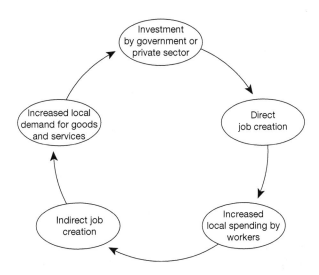

Figure 2.2 Multiplier effect.

to promote investment either through monetary policies such as changing interest rates, or directly through government expenditure. For example, if a government funds a road-building scheme, this will create jobs not only for the road builders, but also for suppliers of road-building materials and tools. The workers will spend money, so supporting other people's jobs, and companies will make profits which can be invested further in productive capital. Government expenditure was, therefore, viewed as a way of 'taming crises' (Preston 1996: 157). This approach was adopted by US President Franklin D. Roosevelt in his 'New Deal' policies of the 1930s in an attempt to boost the US economy and job creation in the aftermath of the Wall Street Crash.

While Keynes did not write specifically on the economic conditions of Southern countries, his ideas regarding government activities were drawn on in the post-war period of development interventions. Policy measures to address the effects of the global economic crisis from 2007 onwards have sometimes drawn on Keynesian ideas, and have therefore been termed 'neo-Keynesianist' (see later in this chapter).

Post-war reconstruction

The post-Second World War period gave Western nations the opportunity to consider the most appropriate forms of international organization and intervention to ensure that the economic crises of the 1930s could never happen again. In addition, they wanted to promote a more peaceful world where warfare could be replaced with diplomacy and negotiation. In the sphere of economics, the 1944 Bretton Woods Conference in New Hampshire, USA led to the creation of three key international institutions aimed at promoting stable economic growth within a capitalist system; the International Monetary Fund (IMF), the World Bank and the General Agreement on Tariffs and Trade (GATT) (see Box 2.1). There were 44 countries represented at Bretton Woods. Although decisions were made which were to influence the whole of the non-communist world, the countries represented were largely from the industrialized world. The nature of representation within global institutions will be discussed further in Chapter 7.

The important role of government, or in the case of the Bretton Woods institutions, multilateral organizations, in economic intervention for development was clearly reflected in the Marshall

Box 2.1

Bretton Woods institutions

All of the organizations are part, in theory, of the United Nations system, but in practice they are autonomous.

International Monetary Fund (IMF)   The original aim of the IMF was to maintain currency stability and develop world trade. It continues to do this largely through the provision of support and advice to countries in difficulty. For example, the IMF has been a key institution in the attempts to achieve economic stability after the debt crises of the 1980s and 1990s, the aftermath of the 1997 Asian economic crisis and the global economic crisis from 2008 onwards. As with many international agencies, support for poverty reduction has also become part of its remit.

World Bank Group   The term World Bank is usually used to imply one organization, but in fact there are five agencies within the World Bank Group. Of these, the term 'World Bank' is most correctly applied only to the IBRD and the IDA.

International Bank for Reconstruction and Development (IBRD)   The IBRD was set up in 1945. Its original aim was to assist in the rebuilding and development of Europe, but following the success of the Marshall Plan (see Box 2.2) the attention of the organization turned to the poorer regions of the world. The IBRD provides loans to national governments of middle income and creditworthy poorer countries at below commercial interest rates. Most of its funds are generated through the world's financial markets. As of July 2010 it had 187 members.

International Development Association (IDA)   For the very poorest nations, even interest rates below commercial levels are too high. The IDA was set up in 1960 and provides interest-free loans to the world's 79 poorest countries. Countries have to pay some administration costs, but the IDA provides access to finance that would otherwise be unavailable. Repayment periods are as long as 35–40 years. As of July 2010 it had 170 members and had lent US$207 billion since 1960. It raises most of its money from contributions from its richer country members, although it also receives funds from the IBRD and early repayments.

International Finance Corporation (IFC)   Since 1965 the IFC has worked with private sector companies in developing countries to help improve investment levels and business success. By providing loans, business advice and financial guarantees, the IFC helps reduce commercial risks for the private sector.

Multilateral Investment Guarantee Agency (MIGA)   Investment in the world's poorer countries is often regarded with suspicion by private sector companies, worried about political instability, restrictions on financial movements and the threat of government expropriation of property. To help the flow of private investment, the MIGA provides a guarantees service. This means that if private sector companies do incur losses for non-commercial reasons, the MIGA will cover those losses. This encourages private sector investment. Since it was established in 1988 MIGA funds have guaranteed over US$21 billion of investment.

The International Centre for Settlement of Investment Disputes (ICSID)
The ICSID was set up in 1966 and, like the IFC and MIGA, it seeks to increase the flow of private capital into poorer countries of the world. The Centre provides arbitration in investment disputes and helps promote confidence in dealings between governments and foreign investors.

General Agreement on Tariffs and Trade (GATT)   Set up in 1947, the role of GATT was to promote free trade between its members. Originally consisting of 23 members, by the time the World Trade Organization (WTO) replaced GATT in 1995 there were 124 members. GATT worked through a series of negotiations or 'rounds'. During each round, members sought to make agreements about reducing tariffs and extending free trade to new economic sectors. There were eight rounds during the life of GATT.

Sources: Dicken (2007); IMF (2010a); MIGA (2010); Potter *et al.* (2008); Power (2003); Thomas and Allen (2000); World Bank (2010a, c)

Plan, officially titled the 'European Recovery Program'. This was a programme through which aid was channelled from the USA to fund reconstruction in Europe (see Box 2.2). The programme ran from 1948 to 1952 and reflected Keynesian theory in that investment into infrastructure programmes was not just to recreate physical capital in Europe, it was also meant to contribute to kick-starting the national economies of the region. In addition, the US government felt that contributing to this reconstruction would reduce the possibilities of European nations 'succumbing' to Communism.

The role of the USA as a key actor in international reconstruction and development was also reinforced by US President Harry S. Truman in his inaugural speech in January 1949. The theme of the speech is similar to the philosophy underlying the Marshall Plan; poverty and low levels of economic development in other parts of the world are detrimental not only to people living in these conditions, but also to the peace and prosperity of the USA and other more economically developed countries. Truman argued that the USA

Box 2.2

Marshall Plan

The Marshall Plan, also known as the European Recovery Program, was announced on 5 June 1947 by the US Secretary of State George C. Marshall. Under this programme, the US government provided financial assistance to the governments of Western Europe to assist in rebuilding their infrastructure and economies after the Second World War. US motivation for this assistance was more than goodwill. Restored European economies would provide markets for US products and would also contribute to the maintenance of a viable trading system. In addition, given the concerns about the communist threat, the US administration felt that providing this assistance would reduce the likelihood of shifts towards communism within Western Europe.

Between 1948 and 1952 approximately US$17 billion were transferred as part of the Program. At its peak, the transfers represented 2–3 per cent of US GNP; a figure far higher than that for US government development assistance today. Most assistance went to the UK, France, West Germany and Italy. To administer the Program, the Organization for European Economic Cooperation (OEEC) was set up. In 1961 this organization became the Organisation for Economic Co-operation and Development (OECD). Its membership now numbers 32 countries, the original Western European and North American members being joined by Japan, Australia, New Zealand, Finland, Mexico and South Korea. The Czech Republic, Hungary, Poland and the Slovak Republic have also joined following their transition to market-oriented economies. Slovenia became the 32nd member, joining in July 2010.

Sources: Binns (2000); Chenery (1989); OECD (2010); Preston (1996)

---

should use its technological knowledge to assist poorer parts of the world to improve production levels and therefore the state of economic development and living conditions.

This speech is often held up as the starting point of 'development planning'. For example, Arturo Escobar, in his influential 1995 book *Encountering Development: The Making and Unmaking of the Third World*, stresses the importance of this speech to the development of a particular discourse and policy-making aimed at the non-industrialized world. Escobar quotes Truman's perspective:

> For the first time in history humanity possesses the knowledge and the skill to relieve the suffering of these people [the world's poor] I

believe that we should make available to peace-loving peoples the benefits of our store of technical knowledge in order to help them realize their aspirations for a better life . . . . What we envisage is a program of development based on the concepts of democratic fair dealing . . . . Greater production is the key to prosperity and peace. And the key to greater production is a wider and more vigorous application of modern scientific and technical knowledge.

(Truman 1949 in Escobar 1995: 3)

For some, as shall be demonstrated later in this chapter, this sharing of technical know-how from North to South was part of a single path of progress to development and modernization. However, for Escobar and other post-development theorists, it represents a Eurocentric approach which fails to recognize the range of societies in the South and also the needs and requirements of the local populations.

Linear stages theory

In 1960, Walt Rostow published *The Stages of Economic Growth: A Non-Communist Manisfesto*. Rostow wrote mainly about 'economic growth' rather than 'development' per se, but he did make distinctions between 'more developed' and 'less developed' areas (1960: 2) and his 'stages of economic growth' were the route to 'more developed' status. Under Rostow's thinking, there was one path to 'development' with the final stage being termed 'Age of High Mass Consumption'. Thus in goal terms, 'development' was conceived of as a state where the mass of the population could afford to spend large amounts on consumer products, the economy was largely non-agricultural and very much urban-based. Finally, the subtitle of Rostow's book *A Non-Communist Manifesto*, stressed that development was to take place in a capitalist context, rather than a communist one. As a process, 'development' was defined in relation to modernity, and to a move from agricultural societies with 'traditional' cultural practices, to a rational, industrial and service focused economy.

In the introduction to his book Rostow is careful to stress the heterogeneity of experiences:

I cannot emphasize too strongly at the outset, that the stages-of-growth are an arbitrary and limited way of looking at the sequence of modern history; and they are, in no absolute sense, a correct way. They are designed, in fact, to dramatize not merely the uniformities in the

sequence of modernization but also – and equally – the uniqueness of
each nation's experience.

(Rostow 1960: 1)

However, despite this claim, he starts Chapter 2 with 'It is possible
to identify all societies, in their economic dimensions, as lying within
one of five categories' (1960: 4). To highlight the nature of
'development' as a process, Rostow used the analogy of an aeroplane
moving along the runway until it reaches take-off and then soaring
into the sky (see Figure 2.3). To demonstrate that this was a route
that all countries could take, he provided information about when
different countries of the world had reached certain stages (see
Table 2.1). Rostow's work, therefore, fitted into the conception of
development as being modernity to be achieved through following
Western models of 'success'. It is probably the best example of what
has been termed 'modernization theory'. The stages-of-growth model
also conceived of development and policy-making taking place
within state boundaries at a national scale, the assumption being that
this was the appropriate scale at which to practise development.

Structural change models

The concept of gradual shifts over time along a particular path was also
found within what Michael Todaro (2000) terms 'structural change

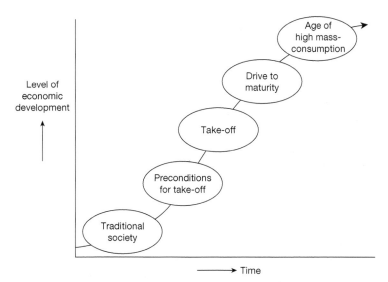

Figure 2.3 Rostow's stages of economic growth.

Table 2.1 Rostow's stages of economic growth

Stage	Characteristic	US Example
Traditional	Based on agriculture; pre-Newtonian science and technology; family and clan allegiances key; pre-nation-state	**Pre-nineteenth century** Native American subsistence and hunter-gatherer societies; European settlers focus on trade of agricultural goods
Preconditions for take-off	Savings and investment rates above population growth rates; national level organization and institutions; new elites; changes often triggered by external intrusion	**1815–40s** Focus on economic activities after independence gained in 1776; higher productivity in agriculture, e.g. cotton production; large-scale infrastructure projects with government funding, e.g. Erie Canal, railway network
Take-off	Stimulus to take-off needed, e.g. political revolution, technical innovation, changing international economic environment; investment and savings rates 5–10 per cent of national income; substantial manufacturing sector; appropriate institutional arrangements, e.g. banking system	**1843–60** American North 'took off' during this period – American South did not 'take off' until the 1930s. Expansion of railways into Mid-West in 1850s associated with inflow of foreign capital; massive expansion in grain exports; growth of manufacturing in East
Drive to maturity	Extended range of technology adopted; development of new sectors; investment and savings 10–20 per cent of national income	**Maturity reached by 1900** Expansion of steel production; agricultural productivity increased; focused economic development policies
Age of mass high consumption	Widespread consumption of durable consumer goods and services; increased spending on welfare services	**1900s onwards** Rise of middle class as move into urban employment in manufacturing, transport, construction; massive increase in consumer purchases, e.g. cars and cigarettes; increased suburbanization; 1913–14: introduction of Ford assembly lines

Source: adapted from Rostow (1960)

models'. The basic theme of these models of development was how national economies shifted from a rural, agricultural base to an urban, manufacturing one. Thus 'development' was conceived of as a largely economic phenomenon.

The key theorist was W. Arthur Lewis, who used his experiences of growing up in St Lucia to examine the nature of economic development (Szeftel 2006). He conceived of the economy of 'underdeveloped countries' being dualistic, i.e. divided into two. The so-called 'traditional' sector consisted largely of subsistence agriculture, although some forms of urban self-employment could also come under this heading. The 'modern' sector, in contrast, was made up of commercial agriculture, plantations, manufacturing and mining. For Lewis, 'development' took place as surplus labour moved from the non-profit oriented traditional sector to the capitalist modern sector. He argued that because there was so much 'surplus' labour, i.e. unemployed or underemployed people, in rural areas, the wages in the modern sector would not increase until the surplus labour had been absorbed (Lewis 1964). Because of the urban basis of much 'modern' economic activity, the Lewis model was based on large amounts of rural–urban migration.

Lewis was concerned with how countries could begin to develop the 'modern' sector. In particular, how countries could raise enough money for investment when the population was saving very little because there were high levels of poverty. As a way out of this trap, he advocated foreign investment. Governments should encourage foreign companies to invest their capital into domestic industrial development through a process termed 'industrialization by invitation' (Lewis 1955).

Lewis' interpretation of development has been criticized from a number of perspectives. Deepak Lal (1985) claimed that his assumptions about non-changing wage rates in situations of surplus labour were untenable, although John Toye (1993: 104) uses the Indian example to demonstrate that Indian wage rates have not gone up a great deal because the continued availability of cheap labour from the 'traditional' sector has helped keep them low. Other criticisms levelled at Lewis have included his failure to recognize the potential contributions that the subsistence agricultural sector can make to economic development (Binns 2008; Toye 1993). His promotion of 'industrialization by invitation' has also been criticized because it encourages dependence (see Chapter 3).

Plate 2.1   Corn stores, West Pokot District, Kenya.
Credit: Katie Willis

Spatial dimensions

As highlighted in Chapter 1, development has not only a social, but a spatial dimension. Just as the Keynesian forms of development policy were meant to lead to a process of trickle-down to the poorest people in society, so development benefits were meant to spread to different regions. For Albert Hirschman (1958), spatially-unbalanced growth was a desirable part of the development process. Based on his own experiences in Latin America he argued that rather than attempt to achieve equal rates of growth throughout a country, it makes sense to allow economic development, particularly industrialization, to be spatially concentrated. His argument was that these so-called 'growth poles' would act as foci for economic development, but that over time, the benefits of such processes would spread and the degree of polarization would reduce. The idea of a route of progression was therefore clear in Hirschman's work. While he recognized that conditions in poorer countries might require different forms of development approach, the underlying commitment to modernization following the Western experience was evident. 'Behind the ideas of unbalanced growth and growth poles it is . . . easy to recognise the conception of growth as a more or less natural and automatic but occasionally disturbed or interrupted process' (Hettne 1995: 43).

The Swedish economist Gunnar Myrdal (1957) also highlighted the spatial inequalities inherent in free market economic development in his book *Economic Theory and Underdeveloped Regions*. However, unlike Hirschman, Myrdal did not believe that spatial polarization would automatically be reversed once economic development reached a certain level. Using the concept of cumulative causation, he argued that once a region started to grow economically, people, resources and finance would be drawn to that area so contributing to further growth. These flows would leave other areas depleted of dynamic people and resources to contribute to development; this is what Myrdal called 'backwash effects'. He did recognize that some benefits of this spatial focus may extend to neighbouring regions through 'spread effects', but overall the vicious cycle of decline for areas outside the core areas would continue.

For Myrdal, the only way in which the exacerbation of spatial inequalities could be reduced was through state intervention. He argued that if state planning was efficient, there was no need for the regional variations in economic growth rates. However, he was aware that in many situations the government and state departments in many countries were not able to achieve this (Myrdal 1970). He termed such states 'soft states' and advocated a move to 'strong states' to ensure that the planning mechanisms could be implemented. However, he did not provide details on how this was to be achieved. Myrdal's faith in planning as a solution to 'development problems' fits with what the post-development theorists would call a Eurocentric technocratic approach. While Myrdal clearly criticized leaving 'development' to the free market, he remained within a development approach that focused on economic growth.

International aid

Flows of international aid from North to South were part of the development policy responses to modernization theory in the post-war period. The poorer countries of the world were regarded as lagging behind on the path of economic development, being largely agricultural and lacking the autonomous capacity for investment and economic growth. Following this interpretation, large transfers of money, technology and expertise were expected to fill the gaps and help the economic development process. Truman's speech about technology transfer fits within this model.

'Aid' can be defined in a number of ways. It usually refers to 'a transfer of resources on concessional terms – on terms that are more generous or "softer" than loans obtainable in the world's capital markets' (Cassen and associates 1994: 2). These resources are usually transferred from one government to another directly (bilateral aid), or from one government through a multilateral agency or an NGO to governments or groups in poorer countries (Lancaster 1999).

In terms of official reporting of international aid flows, data from the Development Assistance Committee (DAC) of the OECD has often been used to assess the level of 'official development assistance' (ODA). To be classified as ODA, flows must be: 'provided by official agencies, including state and local governments, or by their executive agencies', the main objective must be 'the promotion of the economic development and welfare of developing countries' and it must be 'concessional in character and [include] a grant element of at least 25 per cent' (OECD 2008: 1). However, the growing importance of the so-called 'non-DAC donors', including China, India and oil-producing Gulf States, has radically changed the nature of international aid flows (see later in this chapter and Chapter 7) (Mawdsley 2010).

Aid can include grants and loans, but it can also cover technical advice, transfer of resources such as equipment or food, and debt cancellation (see later in this chapter). It is usual to distinguish between 'emergency aid' or 'relief aid', which is mobilized at times of natural disaster or war to meet immediate needs, and long-term development aid. It is this development aid that forms the bulk of aid flows and which has been used to promote particular forms of development practice.

In the 1950s and 1960s, in particular, aid was channelled into industrial development and projects to improve agricultural efficiency through the use of technology. There was also a focus on large-scale infrastructure projects, such as dam construction and road building (Plate 2.2). This form of 'top-down development' was advocated because policy-makers believed that this development path had worked in the North so could work elsewhere. It did, however, have

Plate 2.2 Turkwell Dam, Kenya.

Credit: Katie Willis

serious social (Chapters 4 and 5) and environmental (Chapter 6) impacts. The externally-derived nature of this development approach is summed up by General Olusegun Obasanjo, President of Nigeria, 1976–9 and 1999–2007. He stated 'In education and in industrialization, we have used borrowed ideas, utilized borrowed experiences and funds and engaged borrowed hands. In our development programmes and strategies, not much, if anything, is ours' (1987 in Lancaster 1999: 3).

A growing awareness of the constraints to the effective use of aid, as well as debates around the role of the market in development and the growing importance of aid donors outside the Global North, has resulted in new approaches to international development aid, which will be discussed at the end of this chapter.

Neoliberalism

From the previous sections it is clear that for most governments in the Global North, as well as multilateral agencies such as the World Bank and International Monetary Fund, 'development' in the post-war period was to be achieved through variations on Keynesian approaches. The approach was based on government intervention at a national level and foreign assistance in terms of aid on an international scale. This perspective changed during the 1970s when the role of the state was increasingly questioned.

In the 1970s, some theorists began to argue that the widespread involvement of the state in economic activities, was leading to inefficiency and slower rates of economic growth than would be achieved if the market were left to its own devices. These theorists, such as Deepak Lal (1983) and Bela Balassa (1971, 1981), were drawing on the classical economic theories of Adam Smith and others regarding the 'invisible hand of the market'. For neoclassical or neoliberal theorists, the route to greater economic growth and therefore greater levels of well-being for all, was through reducing state intervention and letting the market set prices and wages. It was argued that this would ensure the most efficient allocation of resources so optimizing growth rates with concomitant social benefits. In relation to aid, Peter Bauer (1972) argued that foreign aid also contributed to the inefficiencies and that this form of intervention should be greatly reduced.

John Toye (1993) describes this shift in theorizing about development as a 'counter-revolution'. He summarizes the three main policy approaches which this counter-revolution wanted to challenge:

- the over-extension of the public sector;
- the over-emphasis of economic policies on investment in physical capital, i.e. infrastructure, rather than human capital such as education and health;
- the widespread use of economic controls, such as tariffs, subsidies and quotas, which distorted prices.

(Toye 1993: 70)

All three of these concerns focus on factors which are internal to a country. There is no consideration of external factors which might influence economic success.

For example, Balassa's work concentrated on the liberalization of trade. Using the cases of four Latin American countries and five Asian countries, Balassa (1971) examined the role of the state in promoting economic growth through industrialization behind tariff barriers. For the period 1950–69, he argued that growth rates in Korea and Taiwan were much higher than in the other seven countries, largely because they adopted an outward-oriented manufacturing export strategy (Figure 2.4). They were adopting the non-protectionist approach that Smith had promoted in the eighteenth century (although see later for further discussion of East Asian development).

By 1983 it was clear that the World Bank had taken on this way of thinking about economic development. In the 1983 *World Development Report*, there was a focus on stressing the relationship between economic growth rates and the degree of state intervention in prices. The implication was that the most rapid growth rates were found in those countries which were most outward-oriented and where states were least involved in 'distorting the market'. However, as Toye (1993: 108) highlights, 'only one-third (or 34 per cent) of the economic performance of these countries is explained by policy-induced price distortions of all kinds. Two-thirds of their economic growth responds to other factors.' There is no consideration of the social and political situations and institutions in these countries which may have affected economic growth, rather the World Bank, following a neoliberal interpretation, prefers to focus on government involvement in economy.

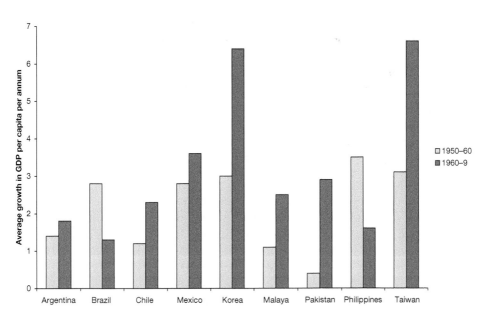

Figure 2.4 GDP per capita growth rates for selected Latin American and Asian countries, 1950–69.

Source: based on data from Balassa (1971: 180)

This move towards a celebration of the market as a neutral resource-allocating system, was found not only in relation to the Global South, but also in the so-called 'Second World' with moves away from state socialism (see Chapter 3) and in the industrialized countries of Western Europe and North America. The United Kingdom under Margaret Thatcher, elected in 1979, and the USA under Ronald Reagan, who took office in 1981, are probably the best examples of this economic philosophy in action.

Debt crisis

The need for an alternative to Keynesian approaches was regarded by some as imperative as economies throughout the world experienced slowing rates of growth in the 1970s. Within many parts of the South, import-substitution policies (see Chapter 3) had been implemented with some success, but the limits of such policies were becoming apparent. These national limits, combined with falling commodity prices and a slow-down in the world economy, led to what has been termed the 'debt crisis' of the 1980s.

The relationship between income coming into a country and that going out is termed the 'balance of payments' (see Table 2.2). These are divided between the current account and the capital accounts. If there is a deficit in one of these accounts, this is not necessarily a problem, as a surplus on the other account may cover the amount. However, if it does not, or if there is a deficit on both accounts, then money must be found to deal with the deficit. This may be found in the cash account, which includes three forms of reserves; foreign 'hard' currencies, such as US dollars, gold and Special Drawing Rights (SDRs) at the IMF. SDRs give the holders the ability to claim currencies from IMF members in the form of loans (IMF 2010c).

The debt crisis arose because many nations of the South were not able to cover their debt repayments (either the interest or the repayment of the amount borrowed). During the 1960s and particularly during the 1970s, many Southern governments borrowed large amounts of money to fund large infrastructure and development schemes. Before the 1970s, most of this borrowing was from Northern governments or from multilateral agencies, but in the 1970s there were increasing levels of borrowing from private banks. This is part of what has been known as the recycling of petrodollars. Because of rising oil prices, countries that were part of OPEC (Organization of Petroleum Exporting Countries) were amassing large amounts of money.

Table 2.2 *Balance of payments accounts*

	In	Out	
Exports of goods and services	A		
Imports of goods and services		B	
Net investment income[a]	C		
Debt-service payments		D	
Net remittances and transfers[b]	E		
Total current account balance (A – B + C – D + E)			**F**
Direct private investment[c]	G		
Foreign loans (public and private), minus amortization[d]	H		
Increase in foreign assets of domestic banking system		I	
Resident capital outflow (capital flight)[e]		J	
Total capital account balance (G + H – I – J)			**K**
Increase (or decrease) in cash reserve account			**L**
Errors and omissions (L – F – K)			**M**

Source: adapted from Todaro (2000: 542–3)

Notes

a The balance between the earnings on overseas investments (such as bank deposits and bonds) which are brought into the country and investment income and profits which are taken out of the country.

b The balance between private transfers into and out of the country, e.g. money sent back home by nationals working overseas.

c This is made up largely of investment by multinational corporations.

d Foreign loans from private banks, governments and multilateral agencies minus the payments made to pay back part of the sum borrowed.

e Movement of money out of the country to foreign bank accounts, property, stocks and shares purchases; triggered by loss of confidence in the domestic economy.

This was deposited in banks and was then lent to other countries so accruing interest. Southern governments were happy to borrow this money to fund their development projects, and this action seemed sensible as interest rates were low and export earnings from commodities remained at a healthy level.

Unfortunately, in the late 1970s commodity prices fell. As Southern countries earned most of their export revenue from primary commodities, either agricultural products such as coffee or sugar, or minerals such as coal or iron ore, a decline in world commodity prices was catastrophic. This fall in commodity prices was exacerbated by global recession in 1981–2, which led to industrialized countries implementing greater forms of protectionism such as increased import tariffs, again making it harder for Southern countries to export their goods. In addition, interest rates went up and millions of dollars of savings were moved by investors to what were

Table 2.3 *Debt burden, 1970–92*

	Total external debt as % GNP			Debt service as % exports of goods and services		
	1970[a]	*1982[a]*	*1992[b]*	*1970[a]*	*1982[a]*	*1992[b]*
Low-income economies	17.0	18.9	32.1	11.3	8.8	18.9
Low-income economies excluding China and India	20.9	28.7	61.2	5.7	9.9	24.5
Lower-middle-income economies	15.4	27.2	40.0	9.2	16.8	17.8
Upper-middle-income economies	10.8	23.2	30.5	10.7	16.9	18.9

Sources:

a   World Bank (1984: Table 16)
b   World Bank (1994: Table 23)

regarded as 'safer' countries through the process of 'capital flight'. These events meant that many Southern governments were no longer able to meet their debt repayments, ushering in the widespread implementation of neoliberal policies based on market-centred theories of development (Milward 2000).

In August 1982 the Mexican government announced that it would not be able to meet the repayments on its debt. This triggered the 'debt crisis', although the problems of debt had been increasing throughout the late 1970s. Debt is not necessarily a problem; being unable to repay the debt or meet interest payments is. In addition, having to spend large amounts of export earnings on debt servicing means there is less to spend on welfare programmes and national investment (Table 2.3).

Structural adjustment programmes

Probably the most well-known aspect of neoliberal development theory in practice has been the implementation of structural adjustment programmes (SAPs) since the late 1970s. These policies were often adopted by national governments in return for continuing financial support from the International Monetary Fund and World Bank. The underpinning philosophy of SAPs reflected the market ideologies adopted by the Thatcher and Reagan administrations and their implementation demonstrates the ways in which policies developed in the North could be imposed on Southern nations.

During the late 1970s and early 1980s, national governments found themselves increasingly unable to pay the interest on the debt they had accrued through borrowing both from commercial banks and multilateral organizations.

SAPs encompassed a series of government-led policies which aimed to reduce (not eliminate) the role of the state in the running of the national economy. SAPs usually included two categories of policies which can be classified as stabilization measures and adjustment measures. The first group included policies such as freezing government-sector wages, cutting back on government expenditure and devaluing the currency. Once the economy had been 'stabilized' adjustment measures were introduced to make longer-term changes which would, it was argued, contribute to a more economically prosperous future. Such measures included opening up the national economy to foreign investment, reforms in the tax system and privatization (Simon 2008) (see Table 2.4). Through these policies, government income would be maximized and there would be much greater efficiency and economic growth. Given the debt burdens and the negative rates of economic growth that had been experienced, such policies appeared to hold some hope for development. These policy recommendations are sometimes referred to as reflecting the

Table 2.4 *Main characteristics of structural adjustment programmes*

Internal policy reforms – to increase role of the market in the domestic economy

- Privatization of state firms – allows for greater competition, reduces drain on state resources if firms doing badly;
- removal of state subsidies – increases competition and reduces state expenditure;
- improvements in tax system – increases state income;
- removal of wage controls, e.g. minimum wages – wage levels should be set by the market;
- reduced government workforce – cuts back on bureaucracy and inefficiency and reduces state expenditure.

External policy reforms – to encourage foreign investment and increasing exports

- Currency devaluation – makes imports more expensive and exports cheaper;
- removal or reduction in tariffs – encourages international trade;
- removal or reduction in quotas, e.g. legal minimum amount of domestically sourced inputs – encourages foreign investment and export;
- end state control of exports, e.g. for agricultural commodities – improves efficiency and encourages private investment.

Source: adapted from Milward (2000)

'Washington Consensus', a term that was originally coined by John Williamson in 1990 to describe the neoliberal policy reforms in Latin America that institutions based in Washington DC, such as the IMF and World Bank, were proposing in the late 1980s. However, it was often used to describe the implementation of neoliberal policies by IFIs throughout the world and as 'a synonym for market fundamentalism' (Williamson 2000: 256).

In the vast majority of cases, SAPs proved to have very serious consequences. The withdrawing of the state, the opening up of the national economy to foreign investment and currency devaluation did not have the desired effect; rather poverty levels increased as real wages went down, unemployment increased and the cost of living rose. The removal of state safety nets in some cases also left the most vulnerable and destitute with no form of assistance (Cornia et al. 1987) (Box 2.3). David Simon (1995: 17) argues that the debt crisis is usually associated with Latin America, but the issues of indebtedness, and in particular the role of SAPs, in Sub-Saharan Africa, should not be ignored. Economic stabilization may have been achieved, but the costs in terms of human welfare have been severe (Simon et al. 1995).

It is difficult to generalize about the effects of SAPs in social terms, partly because of the diversity of experiences (Stewart 1995), but also because we cannot know what would have happened if SAPs had not been introduced. In some cases (Box 2.3) there is evidence of poverty increases after SAP introduction, '[h]owever, the key point is that it is agreed that although SAPs may not have caused poverty in a direct sense, they certainly did not lead to poverty reduction' (McIlwaine 2002: 99).

Growing awareness of the detrimental social effects of SAPs and the move towards a global poverty reduction agenda, led the World Bank and IMF to reshape SAPs in the late 1990s. Neoliberal adjustment policies are still a key part of the organizations' conditions for lending funds, but greater attention is paid to the needs of the poorest people in society. The diversity of national situations is also acknowledged, compared to the one-size-fits-all approach under SAPs. These new policies are grouped under the heading 'Poverty Reduction Strategies' (PRSs) and a key element is the Poverty Reduction Strategy Paper (PRSP), which was introduced at the end of 1999. The PRSP is a document which each government must produce in order to qualify for further funding. Although it is written

Box 2.3

Structural adjustment in Jamaica

During the 1970s the Jamaican economy experienced severe economic problems including negative levels of economic growth from 1974 onwards. These problems led the Jamaican government under Edward Seaga to sign an agreement with the IMF in 1977 for further funding dependent on Jamaica following adjustment policies. Further agreements with both the IMF and World Bank followed in the 1980s and 1990s.

Government expenditure fell dramatically as part of these policies. The budgeted expenditure for 1985/6 was 71 per cent of the 1981/2 level. While government spending on social services was 641 million Jamaican dollars in 1979/80 this had fallen to 372 million in 1985/6 (based on 1979/80 prices). These declines in government social spending, combined with increased unemployment and falling real wages led to declining standards of living. Levels of infant malnutrition increased, education levels fell and the supply of new housing for low-income groups shrank to almost zero. Between 1980 and 1985, O'Level pass rates fell from 62 per cent to 34 per cent, reflecting the declining investment in school infrastructure and teachers' wages, as well as poor levels of health among children and pressures from families for children to enter paid work.

Overall poverty levels increased during the 1980s, with some annual fluctuations. Using US$60 per month at 1989 prices as the poverty line, the poverty rate increased from 45.5 per cent in 1989 to 54.5 per cent in 1996. Some households were able to maintain living standards, especially those who had access to remittances from family members overseas. The role of women in addressing the effects of declining incomes and rising expenses was particularly important (see Chapter 5 for a discussion of the gendered dimensions of SAPs). Beverley Mullings (2009) also identifies the importance of 'gang welfare' in providing support to residents of particular urban neighbourhoods. SAPs in Jamaica have had serious negative impacts on social development in the country, not only in health and education terms, but also with respect to social breakdown.

Sources: adapted from Boyd (1987); Handa and King (1997, 2003); Mullings (2009)

by national governments, the formulation of the report and policy suggestions should be conducted in consultation with civil society organizations and donors (World Bank 2010b). This is to promote a more participatory approach to national development (see Chapter 5). Despite the change of name, many argue that PRSs are little different from SAPs and that while consultation processes have occurred, the quality of participation in terms of giving marginalized groups a voice has been limited (Bradshaw and Linneker 2003; Lazarus 2008).

The 'Asian miracle'

The experiences of the East Asian nations from the 1960s to the 1990s were regarded as evidence of how neoliberal policies could lead to development. Not only did these nations experience high levels of economic growth, but this was associated with improvements in living standards and did not result in increasing levels of economic and social inequality (Figures 2.5 and 2.6). The economic development of the region has been represented as a series of waves. Following the success of the Japanese economy in the post-war period, the newly-industrializing countries or economies (NICs or NIEs) of Hong Kong, Singapore, Taiwan and South Korea experienced rapid economic growth based largely on labour-intensive manufacturing industries. In the 1980s this trajectory was also taken by Indonesia, Thailand and Malaysia, as well as China and Vietnam, which were starting to move away from a state-controlled economy (see Chapter 3).

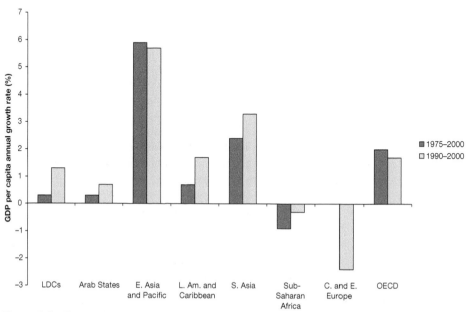

Figure 2.5 Economic growth rates by world region.
Source: based on data from UNDP (2002: 192–3)

High economic growth and low income inequality	High economic growth and high income inequality
China	Botswana
Hong Kong	Gabon
Indonesia	Malaysia
Japan	Mauritius
Singapore	
South Korea	
Taiwan	
Thailand	

Low economic growth and low income inequality	Low economic growth and high income inequality
Austria	Argentina
Australia	Bolivia
Bangladesh	Brazil
Belgium	Chile
France	Colombia
India	Ghana
Italy	Ivory Coast
Malawi	Kenya
Mauritania	Mexico
Nepal	Peru
Pakistan	Philippines
Spain	Sudan
Sri Lanka	Venezuela
Switzerland	Zambia
UK	

Figure 2.6 Economic growth and income inequality, 1965–89.

Source: adapted from Page (1994: Figure 2)

Notes: High economic growth: increase in GDP p.c. of 4% p.a. or above. High income inequality: richest 20% have over 10 times income share of poorest 20%.

In 1993 the World Bank published *The East Asian Miracle*. This book sought to highlight the ways in which economic and social development had been achieved by the East Asian nations by following a few key neoliberal tenets, notably, opening up the economy to foreign investment and trade; limited role of the state in the national economy; investment in human capital, especially education. The book presented this 'East Asian model' as a blueprint for economic development with equity in other regions of the world. As Dixon (1999: 206) highlights: 'the successful Pacific Asian economies have been presented as the living embodiment of neoliberal orthodoxy and examples to be followed by the Third World as a whole.'

Following its publication, numerous commentators queried the validity of the interpretation that the World Bank had made of the post-war economic development in the East Asian countries. Four main criticisms were levelled at the World Bank interpretation. First, that rather than being a coherent set of policies throughout the region, in fact there were a range of approaches adopted within individual countries (Rigg 2003). Second, and very importantly, many criticized the way in which the World Bank used the East Asian experience to promote the benefits of free trade and open economies. In fact, as numerous authors stressed, many of the East Asian countries had achieved their economic success through important government interventions, such as the protectionist policies around Hong Kong's textile industry (Amsden 1994). Third, the World Bank report failed to acknowledge the continued levels of poverty and inequality in the region's nations. Finally, the use of the term 'miracle' suggested an unexplainable process, whereas policies promoting capital accumulation, improving technology and investing in education and training could all be identified as contributing to economic growth (Garnaut 1998).

Plate 2.3 The Petronas Towers, Kuala Lumpur, Malaysia.
Credit: Katie Willis

In 1997 this model of neoliberal economic development collapsed in what was termed the 'Asian Crisis'. The trigger for the economic crisis was the devaluation of the baht, the Thai currency, but there had been cracks appearing throughout the region's economies before this. The crisis was largely financial, in that the withdrawal of large amounts of foreign capital from the region meant that for many countries, the stability of their economies was severely threatened. In 1996, foreign investors' confidence in the Thai economy was shaken by declining export growth and there were also concerns about the large size of the current account deficit. As financial capital was withdrawn from Thailand, investors also lost confidence in other regional economies so capital flight spread (Poon and Perry 1999). This had devastating effects on the levels of economic growth, and also on the levels of poverty among the mass of the populations. Of course, while the crisis spread throughout the region, the experiences were different and some countries weathered the storm much more effectively than others: Indonesia, Thailand and South Korea all experienced negative economic growth rates following the crash, while Taiwan and Japan experienced little change in GDP growth rates (Garnaut 1998).

For the World Bank and other institutions and governments which had held up East Asia as *the* model for successful development policies, the 1997 Asian crisis was certainly a blow. However, World Bank interpretations of the causes of the crisis focused on the failure of the region's governments to follow the neoliberal approach properly. State intervention in the economy which had not been greatly acknowledged in *The East Asian Miracle* was now recognized as a key element in the creation of the crisis (Dixon 1999). Throughout the region, IMF 'rescue packages' were implemented. These were very similar to structural adjustment policies and included conditions such as increased openness to foreign ownership in the banking and financial sectors.

Good governance

The IMF and World Bank, through SAPs, 'rescue packages' and other activities, attach conditions to the loans they provide. This concept of 'aid conditionality' is also part of government aid whether bilaterally or through multilateral organizations, such as the EU. For the IFIs and the DAC donors (see above) the importance of 'good governance' has increased since the 1990s (Weiss 2000).

'Good governance' originally focused on the operation of the public sector and fitted very well within the neoliberal policies of reducing state involvement in the economy. This still remains important and includes processes such as civil service reform, improvements in budgetary management and building capacity of government staff (Harrison 2005). It also attempted to make the accounting process as transparent as possible to reduce corruption and the misappropriation of aid funds. While this transparency was primarily part of making governments accountable to donors, IFIs argued that governments should also be more accountable to their populations. Thus good governance reforms would make government bodies more accessible and responsive to the population's demands. This is part of the growing focus on participatory development (see Chapter 4).

From an original focus on reforming government activities as part of neoliberal economic restructuring, 'good governance' has expanded to include broader themes such as multi-party elections, freedom of the press and respect for human rights. Following the end of the Cold War (see Chapter 3), Northern governments no longer felt the need to support 'undemocractic' governments on geopolitical grounds. The World Bank has six governance indicators, including voice and accountability, political stability and the absence of violence and government effectiveness (World Bank 2010f).

The importance of these good governance indicators can be seen in the conditions applied to countries eligible for debt relief under the Heavily Indebted Poor Countries (HIPC) initiative. This was launched in 1996 by the IMF and the IDA (see Box 2.1). In return for meeting certain conditions, including the PRSP process and appropriate governance reforms, the debt burden of the economically poorest countries of the world is reduced. As of August 2010, of the 40 potentially eligible countries, 28 were receiving support (HIPC 2010).

The HIPC initiative represents a continuation of development policies implemented by Northern-based IFIs in Southern countries. However, development assistance and aid is increasingly coming from non-Northern sources with the rise of the BRICs and the oil-producing states of the Middle East. For example, Brazil's foreign aid programme included US$1,200m in bilateral and multilateral aid in 2010 and US$3,300m in loans between 2008 and the first quarter

Plate 2.4   UNICEF-funded school project, Mulanje, Malawi.
Credit: Katie Willis

of 2010 (*The Economist*, 17 July 2010a). While Brazil's aid tends to be focused on social programmes and agriculture, Chinese assistance is channelled into infrastructure projects. There are usually conditions attached to all development assistance, regardless of source, but the 'good governance' is often not part of non-DAC donor conditions (Mawdsley 2008; Six 2009).

Global economic crisis and neo-Keynesianism

Just as the Asian financial crisis of 1997 challenged prevailing ideas about neoliberalism, so the global economic crisis of 2007 onwards has generated much debate about the nature of neoliberalism and appropriate strategies to achieve sustained economic growth. The crisis was triggered by a collapse in the US sub-prime mortgage market, but soon spread throughout the world because of the way in which financial products were sold and resold as part of global financial transactions (Aalbers 2009). Global trade fell as national economies contracted, leading to increases in unemployment and further reduction in demand for goods and services due to the multiplier effect. For example, in May 2009 an estimated 23 million migrant workers in China were unemployed, of whom 14 million had returned to their villages (Branigan 2009).

For Marxist theorists (see Chapter 3), this economic crisis was an inevitable outcome of the operation of neoliberal capitalism (Harvey 2010). However, for those supporting a pro-market approach, the crisis reflected poor regulation of the financial system, rather than an inherent problem with neoliberalism. The international response to the crisis focused on improving regulation, but also involved significant additional government expenditure intended to stimulate the economy (Box 2.4). This reflects a form of Keynesian economic policy, so has been termed 'neo-Keynesianism'.

The G20 grouping (19 economically dominant countries and the EU) agreed at summits in Washington in 2008 and London in April 2009 to promote government spending and increased support for the IMF to help countries in difficulty. However, by the June 2010 Toronto G20 summit, divisions within the G20 were apparent, with some countries continuing to support economic stimulus and others promoting cutbacks in government spending to reduce deficits. As of August 2010, the outcomes of such policies are unknown, but this demonstrates the continued debates around the role of government in economic development.

Box 2.4

Economic stimulus in the USA

While there had been problems in the housing markets in the USA in previous years, it was in 2007 that the number of housing repossessions increased greatly. By the end of 2007 about 4 million people had lost their homes because they were unable to pay their mortgages. House prices fell throughout the country and banks were increasingly under threat of collapse because they could not cover their financial liabilities.

The US government responded by launching a series of measures which involved pumping large amounts of federal money into the economy. By the end of 2009 it was estimated that the economic stimulus package amounted to US$1.7 trillion. This included US$600 billion to support the financial institutions. It was argued that the government could not allow all the Wall Street banks to collapse as this would be catastrophic for the US economy. Critics argued that the banks should be allowed to fail as the financial crisis was a result of bank misconduct.

In addition to the 'banking bailout', the Bush administration spent US$170 billion in 2008 and the Obama administration spent US$780 billion in 2009 in an attempt to encourage spending, job creation and overall economic growth. Policies included subsidies for homebuyers and tax refunds. Despite this spending, unemployment in 2010 remained at around 10 per cent, but it is not known what levels it would have reached without this additional money in the economy.

Source: adapted from *The Guardian* (28 July 2010, 30 July 2010) (Harvey 2010)

Summary

- Classical and neoclassical economic theories stress the importance of the free market for development.
- Modernization theories argue that development is largely economic and the same development path should be followed by all countries.
- Post-Second World War, Northern countries focused on providing assistance to Southern countries to help them follow the same development path taken by the North.
- Neoliberal policies involving reduced state involvement and a greater role for the market have dominated international development policy since the 1980s.

● Good governance has become an increasingly important condition of international development assistance and explanation for successful development policies.

Discussion questions

1 According to classical economic theories, what are the benefits of free trade?

2 What are the stages of Rostow's linear stages theory and how do countries move from one stage to another?

3 What role did Keynes think the state had in promoting economic development?

4 How and why have neoliberal theories come to dominate international development practice?

5 What is 'good governance' and what role does it play in development theory and policy?

Further reading

Harvey, D. (2007) *A Brief History of Neoliberalism*, Oxford: Oxford University Press. A very clear discussion of the rise of neoliberalism and its implementation in both Northern and Southern countries.

Mohan, G., E. Brown, B. Milward and A.B. Zack-Williams (2000) *Structural Adjustment: Theory, Practice and Impacts*, London: Routledge. An accessible account of the background to structural adjustment, what it involves and the political, economic, social and environmental impacts. There is also a consideration of alternatives to structural adjustment.

Toye, J. (1993) *Dilemmas of Development*, 2nd edition, Oxford: Blackwell. A very detailed examination of the shift from Keynesian approaches to economic development to neoliberal ones. Some students may find the economic content rather hard-going at times, but the book is an excellent overview of these debates.

World Bank (1993) *The East Asian Miracle*, Oxford: Oxford University Press. Excellent example of IFI approaches to economic development in the late 1980s and early 1990s.

Useful websites

www.adamsmith.org   The Adam Smith Institute, which aims to increase awareness of the work of Adam Smith and the role of market-led economic development.

www.brettonwoods.org   Bretton Woods Committee. Provides information about international financial matters and the role of the IMF, World Bank and WTO in global economic policy.

www.g20.org   Website of the G20. It includes information about G20 activities in relation to the global economic crisis.

www.imf.org   International Monetary Fund. Provides information on the approaches and activities of the IMF.

www.londonsummit.gov.uk   G20 London Summit website. Provides information about agreements regarding a global plan for economic recovery.

www.oecd.org   Organisation for Economic Co-operation and Development. Provides information about the organization's activities, including official development assistance.

www.usaid.gov   United States Agency for International Development.

www.worldbank.org   World Bank. Details the activities of all members of the World Bank Group.

www.worldbank.org/wbi/governance   World Bank governance and anti-corruption website.

www.worldbank.org/hipc   Highly Indebted Poor Countries (HIPC) initiative.

www.worldbank.org/poverty/strategies   Poverty reduction and equity site of the World Bank. Provides information about PRSPs as well as the broader poverty alleviation policies adopted by the World Bank.

3 Structuralism, neo-Marxism and socialism

- Marxist theories of development
- Structuralism and dependency theories
- Socialism and the Soviet model
- Maoism and development in China
- African socialism and Afro-Marxism

The state has an important role to play in all approaches to 'development'. This may be providing a system of regulation, law and order within which the market can operate efficiently, as the theories discussed in the previous chapter argued, or the state can be far more interventionist in the workings of the economy. In the previous chapter we considered the role of the state within Keynesian approaches to development and economic growth. In this chapter the focus will be on theories which have been used to justify much greater state involvement.

Marxist theories of development

Karl Marx's theory of development or progress bore similarities to that of the linear stages models described in the previous chapter. For example, in the introduction to the first volume of *Capital* he stated, 'The country that is more developed industrially only shows to the less developed, the image of its own future' (Marx 1909: xvii). However, the end point, while also having similarities in the focus on urban and industrial life, was very different in social and political terms. Capitalism was regarded as just one stage in a transition; pre-capitalist societies, which Marx differentiated as 'Asiatic', 'ancient' or 'feudal', would be replaced by capitalism, which would be usurped by socialism. Under a socialist or communist regime there would be communal ownership rather than private property and individuals would work according to their abilities and would be provided with according to their needs (Table 3.1). While Marx's ideas have often been interpreted as describing the transition that would happen in all societies, he recognized the European bias of his work and accepted that the process may be different outside Europe (Sheppard *et al.* 2009: 61–3).

The central feature of Marx's analysis was the relationship between capital and labour. The stages outlined above represented different 'modes of production', in that at each stage there was a different combination of 'means of production' and 'relations of production' (Box 3.1). Under pre-capitalist forms of production, individuals worked to provide for themselves and their families, often through subsistence agriculture. Thus, the amount of work or labour completed was just enough to provide food, shelter, clothing and other necessities.

Under capitalism, Marx argued that this relationship shifts. As technology advances and humans' abilities to exploit and use natural resources becomes more effective, more complex forms of organization are possible. For Marx, capitalism is characterized by

Table 3.1 *Marxist stages of social development*

Stage	Characteristics
Ancient/primitive communism, Feudalism or Asiatic	Ancient tribal societies; communal ownership of land, tools and other basic economic resources
	Feudalism: found in 'Western' societies; based on agricultural production organized around large estates; land owned by a few, but tenants able to keep their produce once they had paid the landlord
	Asiatic: found in 'Eastern' societies, e.g. India, China, Turkey, Persia; different classes dominated the economy and the state apparatus; needed to ensure centralized control of important technologies such as irrigation systems
Capitalism	Society divided into those who own the means of production and those who do not; those who do not have to earn a living by selling their labour; key role of the market in allocating resources
Socialism/Communism	Communal ownership of means of production by state (socialism) or the people (communism); industrialization means that people no longer have to struggle for a living and individual needs can be met by the distribution systems of the collective socialism; viewed as a transitional stage to communism

Sources: adapted from Mitchell (2009); Watts (2009); Worsley (1990)

Box 3.1

Definitions of Marxist terms

Means of production   The 'things' which are needed for people to produce goods. These include tools and equipment, as well as land, crops and mineral reserves.

Relations of production   The division of labour, i.e. who does what in the production process. Also includes who decides what is produced and how it is produced, so includes the possibility of unequal decision-making and power. This can be based on who owns the means of production.

Mode of production   The system of social relations organizing production. This includes the relations of production, as well as the state apparatus and the legal system. It also includes cultural norms and ideologies about the way society should work.

Sources: adapted from Peet and Hartwick (2009); Worsley (1990)

---

two major divisions within society, the 'bourgeoisie' who own the means of production, and the 'proletariat' who do not. The only way that the proletariat can survive is to sell their labour, i.e. to work for a wage. This work, however, is not the same form of work as in a pre-capitalist system. Not only do workers labour so as to be able to provide the basics for their families, but they also work to provide for the bourgeoisie. This 'surplus' (the amount of work beyond that needed to meet basic needs) creates profit for the owners because workers are not paid the full value of the goods they produce. This profit can then be reinvested into more factories or land, so creating the conditions for further wealth-generation, but the wealth remains with the bourgeoisie.

Marx viewed capitalism as a necessary stage in the progression towards socialism and also considered it much better than pre-capitalist societies which he characterized as irrational and backward. He interpreted capitalism as being inherently unstable and vulnerable to crises. Eventually, he believed that capitalism would be overthrown and socialist forms of organization and production would prevail.

Marx's focus was European development in the nineteenth century. However, Marx's ideas were expanded and applied to other parts of the world. In relation to the growth of European empires and colonies in Africa and Asia, as well as continued relations of exploitation with independent Latin American countries, Marxist theories of imperialism focused on the way these different parts of the world helped defuse or delay crises in capitalist development in Europe (Peet and Hartwick 2009). For example, Lenin, the leader of the 1917 Revolution in Russia (see later), argued that imperialism was the 'highest stage of capitalism'. According to Marxist theory, capitalism needs ever increasing opportunities to create profit in order to survive. Colonies provided excellent possibilities for further profit generation, through the creation of new markets, new sources of raw materials, and cheap labour (Webster 1990: 82). According to Lenin, once these possibilities had been exhausted capitalism would collapse.

Neo-Marxism

Classical Marxist theories as described above have been criticized for focusing on the experiences of the societies of Western Europe and assuming that all countries of the world would generally follow the same path of progress and development. However, in the 1950s and 1960s neo-Marxist approaches began to question this interpretation. The experiences of newly-independent states in the Caribbean, Africa and Asia showed that Lenin's ideas about imperialism being the highest stage of capitalism could be challenged. Despite de-colonization, capitalism had not collapsed in these countries (Roxborough 1979).

Paul Baran (1960) drew on Marxist ideas, but applied them to world conditions in the mid-twentieth century, hence the label 'neo-Marxism'. He, along with Paul Sweezy, argued that capitalism was now in a period of 'monopoly capitalism' (Baran and Sweezy 1968). Large companies dominated the world economy and were able to exploit poorer parts of the world. Baran argued that the governments of these poorer economies should intervene and prevent funds that could be used for development being siphoned out of the country as profit. Unfortunately, these governments were either corrupt, or lacked the power to prevent this exploitation. For Baran the only solution to this problem was for countries to leave the world capitalist system in favour of a state-socialist system. It was only by doing this that development would be possible. These ideas were similar to those adopted by the 'dependency theorists' (see pp. 69–72).

Structuralists

So far in this book, there has been a focus on theories of development which originate from and are based on the European experience. These Eurocentric approaches have, however, been challenged by a number of theorists, from a wide range of perspectives. Latin American academics and writers have been an important source of these challenges, with the structuralists being a key group.

The structuralist approach to explaining the nature of Latin American economies and levels of development is associated with the United Nations Economic Commission for Latin America (ECLA or CEPAL in Spanish) which was set up in Santiago, Chile in 1947. While there were similar regional commissions in other parts of the South, the United States was rather reluctant to support a Latin American commission, fearing that this would lead to dissent from the US viewpoint regarding the future of the Americas.

The ECLA executive secretary, Raúl Prebisch, along with many others working in the organization, devised their arguments about development theory and strategy based on the Latin American experience. Prebisch's argument was that the low levels of economic growth and standards of living would not be improved through following the free trade arguments of modernization theorists and others (Prebisch 1959). This was because the global economic structure was very different from that which existed when the European countries experienced their processes of industrialization. According to Prebisch, the global trading system based on principles of free trade acted as an obstacle for Latin American development.

What is key here is that the ECLA structuralists were not arguing that 'development' as a goal was not represented by industry, urbanization and other symbols of modernity. Rather they argued that development as a process would be different from the path advocated by Eurocentric theorists. How could the path be the same if the global environment was different? This recognition of the importance of historical context for a consideration of development is similar to the neo-Marxist reworkings of Marxist theory in the light of experiences in the global periphery.

According to the ECLA interpretation, national development strategies should involve greater state intervention to protect national industries, so allowing them to establish themselves without competition from foreign firms. This approach built on the ideas of 'infant industry' developed by a nineteenth-century German economist, Friedrich List. What was termed 'import-substitution industrialization' (ISI) was adopted in a number of countries throughout the region with some initial success. This involved erecting tariff barriers, so that national manufacturing was protected from more efficient foreign firms that would be able to sell their products more cheaply. The addition of a high import tariff meant that the foreign companies' prices were raised, so allowing domestic firms to compete (see Box 3.2).

Another element of the ECLA approach was land reform. Through much of Latin America and the Caribbean land-holding patterns during the colonial period and then into independence had been characterized by massive inequalities. A small number of landholders held vast swathes of land, often termed *latifundia*. This land was often used for plantations or large-scale cattle ranching depending on the location. The majority of agriculturalists were small-scale peasant producers, scraping a living on very small plots of marginal land, known as *minifundia* (Kay 2004; Thorpe and Bennett 2002). Land reform usually involved redistributing land so that these small-scale farmers would have larger plots. This would enable them to increase their production and so contribute to poverty alleviation in their

---

Box 3.2

Import-substitution industrialization in Brazil

Brazil adopted a policy of import-substitution industrialization (ISI) from the 1940s onwards in an attempt to increase levels of economic growth by protecting domestic industry and promoting production for export. The Brazilian government used tariff barriers to reduce imports and encouraged production for export through the use of subsidies.

Between 1965 and 1973 the average annual growth in manufacturing production was 12 per cent. Manufactured goods also increased from 8 per cent of exports in 1965 to 39 per cent in 1982. While government policy was important in this, it should also be remembered that Brazil is the most populous country in Latin America. The large population provided a market for the manufactured goods. For smaller Latin American countries, this option was not available.

Source: adapted from Gwynne (1996)

communities, but also to national economic development through increasing productivity.

The fact that the structuralists were arguing for a capitalist-based form of development suitable for Latin America meant that this approach was not regarded as challenging the capitalist industrialized nations. A key indicator of its acceptance was the incorporation of some of the ECLA policies into the Alliance for Progress programme of the 1960s (Clarke 2002) (see Box 3.3). Land reform was a particularly important part of the Alliance for Progress agenda. However, in most countries, apart from Cuba and Nicaragua where there was revolutionary change (see pp. 81–8), agrarian reform was rather limited. Often peasants were given 'new' land within settlement projects, rather than land that had been expropriated from large landholders.

Despite some successes, the influence of the structuralists on policy implementation declined. This was largely as a result of the

---

Box 3.3

Alliance for Progress

The Alliance for Progress (AFP) was a US programme targeted at social and economic development in Latin America. While the basis for the programme had been developed under the Eisenhower administration, it was only implemented under President Kennedy in 1961. Under the AFP, individual Latin American nations had to present a 'development plan' to a panel of US economic and technical experts. Funding available for the entire scheme was up to US$20 billion over ten years, and the US was also committed to promoting multilateral and private investment in the region.

Another strand to US Latin American policy during this time was an expanded counter-insurgency programme. Both this and the AFP were coordinated by the US Agency for International Development (USAID). The counter-insurgency projects were classified as being for 'public safety'.

While there were some successes under the AFP, by the end of the 1960s it was clear that its goals of social and economic progress following a US liberal model had failed. For example, in 1964 the Brazilian government was overthrown by the military, but Brazil continued to receive funding from the US.

Source: adapted from Skidmore and Smith (2004)

perceived limits to ISI, agrarian reform and other forms of state intervention policies at a national level. Protecting domestic infant industries from foreign competition enabled large numbers of firms to be established, but obstacles to continued progress included limited national demand because of low incomes and the need to import machinery and high-tech equipment as the production process became more complex. These limits were becoming increasingly apparent in the late 1960s and with the oil crises of the 1970s it was clear that some changes had to be made to the national development strategy. As discussed in Chapter 2, these changes were very radical and involved a shift to export-oriented industrialization, less state involvement and opening up to foreign investment. ISI was interpreted as fostering inefficiencies in the operation of the economy and stifling growth and development.

Dependency theories

Another key Latin American theoretical development was 'dependency theory'. Despite the name, the approaches to 'dependency' were rather diverse, so 'dependency theories' in the plural is more appropriate. In addition, some critics of the dependency school (see below) claimed that it was not really a theory.

The key argument of dependency theorists, or *dependistas* as they were known, was that Latin American countries found themselves in positions of 'underdevelopment' because of the operation of the capitalist system. In particular, the core industrialized countries were experiencing growth and economic development through the exploitation of the non-industrialized peripheral countries. The argument differed greatly, therefore, from that of modernization theorists and classical Marxist theorists who saw non-industrialized countries as merely being further behind on the development ladder. According to the dependency theorists, Latin America's development situation was a result of capitalist development, just as industrialization in the North was a result of this process. André Gunder Frank (1967) termed this the 'development of underdevelopment'.

Frank used the examples of Chile and Brazil to demonstrate the chains of dependency that had existed since the colonial period beginning in the sixteenth century. He argued that with capitalist development (he defined capitalism as being production for market exchange) Latin America was caught up in a global system of dependence consisting of relationships of exploitation from the global scale to the inter-personal (see Figure 3.1). Thus, individual peasants were exploited by local land-owners who did not pay them the full value of the commodities they produced. These land-owners then sold the goods to merchants in the urban areas at a higher price than that paid to the peasants, so generating a profit. This chain of exchange and exploitation continued until the surplus generated through these exchanges was taken out of the country to the core. Periods when Latin America was less engaged with the global economy, for example during the Second World War, were, Frank argued, periods when development was most likely to take place within the region. Celso Furtado presented a similar argument in his book *Economic Development in Latin America* (1976).

While dependency theorists would agree with the claims that exogenous (i.e. outside the country) factors were key in explaining the low levels of economic development in Latin America, the solution to this limiting situation differed. Colin Clarke (2002) highlights the main distinctions between the 'reformists' and the 'Marxists'. The former felt, like the ECLA structuralists, that what was needed was reform of the capitalist trade system, perhaps with greater state intervention (see Furtado 1976). The Marxists (or more accurately neo-Marxists) saw the overthrow of the capitalist system as the only solution. Frank was one of the most vociferous advocates of this approach, believing that within capitalism the peripheral regions of the world would always be exploited and marginalized.

Although dependency theory was largely applied to Latin America, the concepts of dependency were also applied to other parts of the world. For example, Walter Rodney's book *How Europe Underdeveloped Africa* (1981 [1972]) argued that the intervention of European powers in African social, economic and political processes throughout the nineteenth century created a situation of dependency and led to the impoverishment of African peoples. Samir Amin (1974) makes a similar argument in the African case, focusing on economic processes, particularly the extraction of primary products.

The fact that the dependency ideas came from the experiences of the Global South was certainly a welcome change from the dominance of Northern voices in theories of development. However, despite the dependency approach's popularity in some circles in the 1970s, its

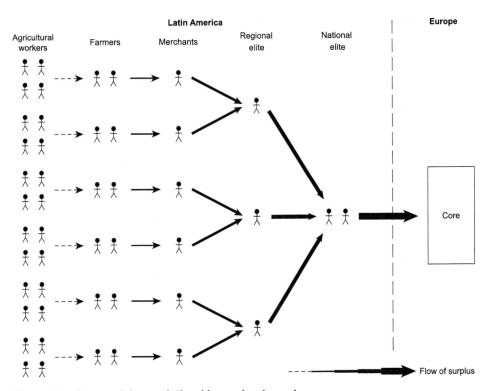

Figure 3.1 Core–periphery relationships under dependency.

influence on policy-making was limited and it has been increasingly criticized. These criticisms come both from empirical evidence which challenges the claims of the dependency theorists, and also from those who query the assumptions on which the dependency approach is grounded.

In relation to empirical support, while the interpretations of Frank, Furtado and others may have had a significant basis in Latin American historical experiences, the conclusions that capitalist-style development was impossible for peripheral countries within the existing system was challenged by the economic success of the newly-industrializing countries of Asia during the 1970s (see Chapter 2). The existence of such evidence refuting the basic claims of the dependency approach undermined the dependency interpretations.

In addition, dependency theories were criticized for being overly concerned with economic factors, without any consideration of the social, cultural or political contexts within which development (or underdevelopment) took place. While dependency theorists had, unlike the modernization theorists, taken a historical view of development by considering *when* processes were taking place, they did not consider the wider contexts within which development occurred. David Booth (1985) also criticizes the dependency school, particularly Frank, for the definition of capitalism used. In Frank's work, capitalist development was defined as 'autonomous industrial growth'. Booth stresses that if this definition is used then it is inevitable that capitalist development will be viewed to have taken place most successfully when ties to the global economy are weaker. According to Booth, this form of circular argument discredits the dependency approach.

World-systems theory

The importance of the global economic system and hierarchies within it was also a key factor in world-systems theory. This was developed by Immanuel Wallerstein (1974) and shares many characteristics with the dependency school. For example, both approaches stress the importance of considering national economic development within a global context, rather than just concentrating on individual countries. The relative strength of states within this global system helps influence levels of development. Both also have a strong historical basis.

However, Wallerstein was keen to move beyond the static dualism of the dependency models. Rather than viewing the world in terms of 'core' and 'periphery', Wallerstein identified three groupings of countries: 'core', 'semi-periphery' and 'periphery'. In addition, the members of these categories were not fixed; over time countries were able to move in and out of categories depending on their economic situation. The inclusion of the 'semi-periphery' was a reflection of global events in the late 1960s and early 1970s. While dependency theorists were arguing that countries in the global periphery were doomed to be forever exploited and marginal, some countries of the world were experiencing economic development in terms of industrialization. These newly-industrializing countries (NICs) included the 'Asian Tigers' of South Korea, Hong Kong, Singapore and Taiwan, as well as Latin American nations such as Brazil.

As with Frank, Wallerstein considered the global capitalist system to date from the fifteenth and sixteenth centuries, when European influence, both economic and political, was expanding beyond the

European heartland. Before the Industrial Revolution in the eighteenth century, European powers competed for dominance, with some countries losing prominence and becoming semi-peripheral (such as Spain) while the countries of north-west Europe became the core. Peripheral regions at this time included those of South and Central America. With industrial expansion in Europe and later in the USA, the core expanded, some nations in the periphery became semi-peripheral and the periphery grew as parts of Asia and Africa were incorporated into the global economic system through processes of colonialism (Peet and Hartwick 2009). In the first decade of the twenty-first century, Sheppard *et al.* (2009) classify the core countries as mainly Western European states, Russia and some Eastern European states, the USA and Canada, Australia, New Zealand, Japan and Israel. The semi-periphery includes some Latin American states, China, India, Malaysia, Turkey, Saudi Arabia and the United Arab Emirates. Within Africa, only South Africa is part of the semi-periphery, while the rest of the continent is classified as being peripheral in the global economic system (Figure 3.2).

The fluidity of categories and the potential movement from one category to another differs from the classical Marxist developmentalist viewpoint outlined at the start of this chapter. Countries do not follow a linear pathway of progress, rather at different times as the global economy changes, certain countries may be able to make economic advances, while others lose out. This historical approach has great benefits, but, as Thomas Klak (2008) points out, the possibilities of testing the world-systems theory are very limited. Rather than helping us explain changing patterns of economic development, the world-systems theory can be used to describe certain patterns, and would perhaps be better described as 'a world-system *approach, analysis* or *perspective*' (Klak 2008: 105, emphasis in the original). As with dependency approaches, Wallerstein's ideas can also be criticized for their focus on state-level action, so excluding local-level processes.

Socialist approaches to development

For some theorists, the only way for development to be achieved was to break from the capitalist-led path of development to an alternative route, albeit with a similar conception of modernity. As stated in the earlier discussion of Marxist interpretations of the workings of

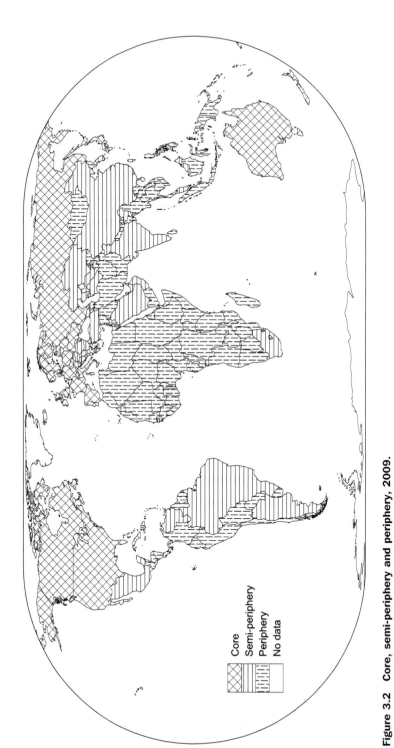

Figure 3.2 Core, semi-periphery and periphery, 2009.

Source: adapted from Sheppard et al. (2009: Plate 2.1)

Map data © Maps in Minutes™ (1996)

society, under capitalism society is divided into two classes; those who own the means of production and those who do not. Under a socialist system, the means of production are owned by the state. Because of this it is argued that profit is no longer the driving force of the economy; rather, the needs of the population are prioritized. Because of state ownership of land, factories, etc., decisions can be made about resource allocation that are made on the basis of need, rather than ability to pay. The role of the state in all aspects of economic, social and political life means that this form of approach is sometimes called 'centrally-planned'.

While socialism has been implemented and experienced in a variety of different ways, there are some common characteristics (Box 3.4). In the mid-1980s, a significant number of countries were experimenting with a socialist form of development. As well as the Soviet bloc countries of the USSR and the states of Eastern Europe, there were large numbers of developing world countries that could be classed as 'socialist' (Table 3.2). In the early twenty-first century the numbers are far fewer and include, most notably China, Vietnam, Cuba and North Korea. However, in all but the North Korean case, economic reform processes are well advanced, including opening up to foreign investment and the loosening of restrictions on private

---

Box 3.4

Characteristics of socialist model of development

Economic
1 state ownership of major industrial enterprises;
2 industrial and infrastructure decisions made according to central government plans, rather than operating through the market;
3 extensive state control over foreign trade and investment;
4 state intervention in the labour market; employment decisions are made according to central plans, rather than market forces;
5 state control of prices;
6 state intervention in agriculture and rural–urban relationships.

Political
1 generally ruled by one party; organized political opposition and many civil society organizations are not tolerated.

Source: adapted from Kilmister (2000: 309)

Table 3.2 *Socialist countries, 1985*

Full member of Comecon	Non-Comecon	Marxist-Leninist	Marginal cases
Bulgaria	Albania (member 1949–61)	Angola	Afghanistan
Cuba	China	Burma (Myanmar)	Algeria
Czechoslovakia	North Korea	Kampuchea	Benin
East Germany	Yugoslavia (associated member)	(Cambodia)	Burkina Faso
Hungary		Ethiopia	Cape Verde
Mongolia		Mozambique	Congo
Poland		Laos	Guinea
Romania		Yemen	Guinea-Bissau
USSR			Guyana
Vietnam			Iraq
			Libya
			Madagascar
			Nicaragua
			São Tomé and Principe
			Seychelles
			Somalia
			Surinam
			Syria
			Tanzania
			Zimbabwe

Sources: adapted from Forbes and Thrift (1987: Figure 1.1) using Wiles (1982) classification system

Notes

Comecon (Council for Mutual Economic Assistance/CMEA): set up in 1949 as a response to the Marshall Plan (see Chapter 2). Organization of socialist states aimed at providing financial and technical assistance (including military help) to members.

Non-Comecon: well-established socialist or communist states, but they remain outside the strong influence of the USSR.

Marxist-Leninist: hard-line socialist governments closely allied to the Soviet bloc, but not Comecon members.

Marginal cases: largely one-party socialist states, but not deeply entrenched socialist systems.

property. Despite the declining popularity of socialist models of development in many parts of the world, a number of Latin American countries have been moving towards more state-led development strategies in opposition to the neoliberal policies advocated by IFIs and Northern governments (see below).

While the basic tenets of the socialist model can be stated, the actual realities of what have been termed 'socialist experiments' very greatly throughout the world. As Andy Kilmister (2000) highlights, socialist forms of political and economic organizing are not introduced into a vacuum. The nature of previous societies and economies will influence outcomes, as will the form of socialist model which is followed. Kilmister also stresses the diversity of routes into socialism. While in the nineteenth century Marx and Engels believed that socialism would be a stage in a linear model coming after the collapse of capitalism, this has proved not to be the case in any of the societies which have followed a socialist path. In the post-Second World War period, Kilmister identifies three main categories of countries which have followed a socialist model:

- Countries where the state was the only actor strong enough to direct development: states such as Afghanistan, Ethiopia and Mongolia remained largely separate from the colonial struggles of the late nineteenth and early twentieth centuries.
- Countries where socialist ideas were key in struggles for national liberation from colonial rule: in many societies seeking to break from colonial rule, socialism provided a key ideology around which to organize. After decades of control from outside forces with policies which created great inequalities in wealth and opportunity, liberation, it was argued, would result in opportunities for all and the nation's resources being owned by the people in the form of state ownership. Countries which adopted this approach included Angola and Mozambique. Kilmister also stresses that the links between socialism and national liberation were also used in response to external forces other than colonial rulers, as in Korea and Vietnam. The Central American and Caribbean nations, in particular Nicaragua and Cuba, adopted a socialist route not to overthrow colonial rulers, but to escape from the tyranny of US-supported authoritarian rule.
- States adopting strong forms of state central planning, but not formally adopting a socialist route: these governments chose to promote economic growth and development in their countries through a very state-centred process. Policy instruments such as five-year plans were often adopted. Good examples include India under Nehru 1947–64, and what has been termed 'African socialism' (see pp. 95–8).

Strong state involvement in welfare services is certainly not only found in countries adopting a socialist model of development, but a commitment to widespread access to health and education services which is part of the socialist model, has had very positive results in a number of countries and regions (see Box 3.5).

Box 3.5

Social development in Kerala, India

In 1957 a Communist government was elected to the southern Indian state of Kerala. Since then, the state has experienced great improvements in social development indicators. Investment in health and education, as well as policies of land distribution and measures to improve women's participation in economic and political activities, have created what some have interpreted as a 'model' for equitable development.

Social indicators are on a par with those for much richer countries. For example, in 2000 the life expectancy at birth was 74 years for women and 68 for men, compared to 64 and 63 respectively for India as a whole (Franke 2002). In the 1950s the crude birth rate was 44 per 1,000, but this had fallen to 18 per 1,000 in 1991 (Sen 1999: 221). While the average fertility rate for India as a whole is 3 children per woman of child-bearing age, the figure for Kerala is 1.7. This can be explained by improved access to contraception and better health care, so that infant mortality rates are now about 12 per 1,000 live births, and improvements in women's literacy levels. Rather than enforcing controls on fertility as in China, the government of Kerala has examined the causes of higher fertility levels and implemented appropriate social policies.

These development policies, while being very successful in social terms, have not been as successful economically. Levels of GDP per capita and economic growth have tended to lag behind those of other Indian states. However, there has been growth in the IT industry and the state continues to attract large numbers of tourists. Many criticisms of the Kerala 'development model' highlight the role of remittances. Thousands of workers migrate from Kerala every year, largely going to the Gulf States. They send back large amounts of money much of which is spent on land, gold or consumer goods. These remittance flows, as well as the ownership structures in the tourist and IT industries, have led to growing levels of inequality in the state. There are also serious concerns about the environmental impacts of uncontrolled development.

Sources: adapted from Franke (2002); Menon (2008); *New Internationalist* (1993); Sen (1999); Tsai (2006); Véron (2001)

In addition, state control of resources, in particular land, factories and infrastructure, can lead to impressive levels of economic development. As both Keynes and Rostow argued within a capitalist setting, for economic growth to take place at certain times there needs to be significant government involvement, perhaps because of the scale of the project, or because of the risks involved. However, as both the Chinese and Soviet examples discussed below demonstrate, these rapid leaps in economic performance may be achieved at great cost to the natural environment (see also Chapter 6), individuals' quality of life, and long-term economic stability.

State control of the means of production, trade and prices means that in theory the economy works for the benefit of the people. As well as providing basic health and education services, socialist practices of development should reduce levels of inequality which can arise when market forces are left to their own devices. In general, inequality levels are low in socialist societies, with Gini coefficients of low to mid-20s. However, it must be noted that in many cases, this is usually within a situation of low levels of per capita income. With the transition to capitalist economies, many ex-socialist states have increased levels of inequality. For example, after transition Russia's Gini index increased 24 points and Lithuania's rose by 14 points between 1987–8 and 1993–5 (Milanovic 1998: 41). There were also increasing levels of unemployment, declines in real income and falling school enrolments (Hörschelmann 2004).

The map of the world's socialist countries today looks very different from the situation in 1985 (see p. 84). The collapse of the Soviet bloc in 1989 represented the most extreme version of what was happening to states adopting socialist models, but throughout the world, economic reforms and in some cases political reforms have led to the questioning of state socialism as a viable development approach. Francis Fukuyama (1989) has termed this 'the end of history'. He states:

> the century that began full of self-confidence in the ultimate triumph of Western liberal democracy seems at its close to be returning to full circle to where it started: not to an 'end of ideology' or a convergence between capitalism and socialism, as earlier predicted, but to an unabashed victory of economic and political liberalism. The triumph of the West, of the Western *idea*, is evident first of all in the total exhaustion of viable systematic alternatives to Western Liberalism.
>
> (Fukuyama 1989: 3, emphasis in the original)

However, events in the first decade of the twenty-first century have suggested that the disappearance of socialism as an alternative development approach has not occurred. The election of left-wing parties in parts of Latin America has provided challenges to the market-led neoliberal hegemony, and the global economic crisis has also resulted in questions about the neoliberal capitalist model (see Chapter 2).

Soviet model of development

The transformation of the USSR from a predominantly peasant society in the 1920s to one of two global superpowers with an extensive industrial sector and urban population by the 1960s, is a process which has attracted a great deal of attention in other parts of the world. The 'socialist experiments' in the Global South drew lessons and inspiration from what went on the USSR.

On the eve of the First World War, agriculture provided over 50 per cent of the national income and about three-quarters of the employment in what was then the Russian Tsarist empire (Davies 1998: 10). Agriculture was dominated by very small-scale peasant production. The October Revolution of 1917 led to the Tsars being overthrown by the Bolsheviks led by Lenin. In Marx's theorizing about the route to socialist societies (see pp. 62–5) the revolution to

Plate 3.1 State housing block, Moscow, 1989.
Credit: Katie Willis

bring about such a society was meant to be led by an urban industrial working class overthrowing a capitalist regime. This was clearly not the situation in the Russian example, so challenging the linear transition ideas of classical Marxism.

The socialist model adopted under Lenin focused on creating an urban industrial economy. These policies were introduced under the New Economic Policy (NEP) in the 1920s (Box 3.6). The industrial focus of this policy reflected Leninist ideas of modernity and a move away from 'traditional' peasant and agriculturally-focused economic policies. Within the economic sphere the state apparatus made decisions regarding what was to be produced, how and where. In addition, in political terms, opposition to the Communist Party which formed the government was increasingly clamped down upon.

In 1929 Josef Stalin came to power and the Soviet project of the social ownership of production intensified. While following the same broad socialist principles as Lenin, Stalinism included the extension of collective ownership, particularly in agriculture, where peasant households were forcibly organized into collective farms. State

---

Box 3.6

New Economic Policy in the USSR

During the 1920s the New Economic Policy (NEP) was adopted under Lenin. This policy was focused on economic development through industrialization. Large-scale industry was state owned and the state also controlled foreign trade. Imports had to be licensed and export earnings were also channelled through the central state. The focus on industrial policy was on capital goods such as machinery and materials such as iron and steel. Concentrating on this form of industrial production, rather than on consumer goods such as clothing, was justified because capital goods contribute directly to increases in production and productivity. Workers in industry benefited greatly from these trends, with improved wages and working conditions. In contrast, during the NEP, agriculture was organized largely through peasant households in village groupings and peasants were able to sell any produce remaining after they had met state targets. Agricultural production remained limited and peasants' standards of living did not improve. A devastating famine in 1927–8 partly reflected this marginalization of the agricultural sector in economic policy.

Sources: adapted from Davies (1998); Kilmister (2000)

control of economic processes increased, as did the exercise of state power, through imprisonment and execution, which was particularly vehement under Stalin. While in the 1920s the one-party Communist state was consolidated and opposition voices were increasingly silenced, under Stalin this intensified; millions of people were imprisoned in a variety of labour camps and prisons, and thousands were executed for supposedly being an enemy of the state. Wheatcroft and Davies (1994 in Davies 1998: 51) estimate that about 682,000 people were executed at the height of Stalin's terror in 1937–8. In January 1953, just before Stalin's death, about 5,223,000 people were interned in camps (Davies 1998: 70).

Under a socialist system not only does the state have great direct influence in the economy, it can also plan social provision based on need, rather than market-led criteria. From the 1920s onwards, there was increased state expenditure on health and education, with very positive results in these dimensions of human development. For example, literacy rates for people aged nine years and over rose from 51 per cent in 1926 to 81 per cent in 1939 (Davies 1998: 46). This reflected both the increase in school enrolment for children, and the success of adult literacy schemes (see also Box 3.5 for similar examples from the Indian state of Kerala).

The centrally-planned economic model with significant investment in social provision, continued until the break-up of the USSR in 1991. By the late 1970s, it had become clear that the economic growth rates achieved under the Stalinist industrialization policies were slowing down. Just as with the limits to ISI described above (pp. 76–7), it was argued that without significant inputs of technology and investment that an opening up of the economy could bring, the Soviet economy would stagnate. In addition, social problems such as alcoholism and crime were increasing as the state was unable to continue supporting the population as it had in the past (Kilmister 2000).

In 1985, Mikhail Gorbachev was elected as the general secretary of the Communist Party and therefore the leader of the USSR. He implemented reform measures termed *perestroika*, which included an enhanced role for the market within resource allocation. For example, enterprises were allowed to make decisions about what and how they produced goods, rather than following centrally-decided targets. This economic reform was also associated with greater political freedom throughout the country and also an attempt at greater openness towards the non-communist world through the *glasnost* approach. Attempts to contain these reforms within the communist USSR system failed. The country broke up and 15 republics became independent, although in 1992 they formed the Commonwealth of Independent States (CIS). The shift to market-oriented economic systems is prevalent throughout the newly-independent republics, and most now operate a form of liberal democracy with regular elections.

In terms of the nature of 'development' during the Soviet period, it is clear that, economically, the drive to modernization was impressive. From a society dominated by peasant agriculture to a highly-industrialized economy, the Soviet experiment provided some examples of how central planning could be used to promote such economic changes, rather than relying on the market. While there are numerous problems with using Soviet economic statistics from this period (see Davies 1998), Western estimates of industrial output suggest rates of increase of about 10 per cent per annum in the period 1928–40, and later rates of approximately 7 per cent were not uncommon. In addition, state provision of education, housing and healthcare benefited millions of Soviet citizens.

However, a centrally-planned economy, especially within a territory of approximately 22 million km², led to a range of inefficiencies and problems. For example, factories were given production targets to meet. As long as the quantity was produced the target was met; little, if any attention was given to quality. The focus on industry meant that agriculture was neglected with serious consequences for food supply. In addition, the Soviet project did not include any recognition of the environmental consequences, leading to very serious environmental damage (see Chapter 6). Finally, limitations on human freedom through the extreme brutality of the Stalinist period, political repression and the outlawing of opposition voices throughout, must be recognized.

Maoism

The Communists led by Mao Tse-Tung came to power in China in 1949. However, while the USSR provided some economic support in the following years, the development strategies adopted by Mao were rather different from those adopted by Lenin and later Stalin. The Chinese revolution was largely a peasant revolution, and the focus on

Plate 3.2   Statue at Mao Mausoleum, Beijing.
Credit: Katie Willis

rural development was paramount during the Maoist period
(Kilmister 2000).

Central state planning was key to the system and a series of five-year
plans were implemented. During the first Five Year Plan (1953–7)
widespread collectivization of agriculture took place, with peasants
grouped into agricultural settlements for communal production.
Despite the rural focus of many policies, the bulk of state investment
went into heavy industry as under the Stalinist model in the USSR
(Hodder 2000). This investment was geared largely towards the iron
and steel industries and also the energy sector in the inland regions
away from the eastern coastal regions (Wu 1987). This geographic
focus was an attempt to spread the benefits of the revolution and to
reduce spatial inequalities. The inland focus was also because of the
perceived need to protect important infrastructure projects from
possible foreign military attack (Wei and Ma 1996).

In the 'Great Leap Forward' of 1958–61 rural development became a
key element to Chinese economic progress. People's communes were
expected to produce agricultural goods, but also some industrial
goods, particularly those, such as machinery and chemical fertilizers,

which would contribute to increasing agricultural productivity. Mao termed this dual focus as 'walking on two legs'. This vigorous attempt at rapid economic development experienced many difficulties. The quality of industrial products was often poor, and the disruption to agriculture caused by the need to spread rural labour efforts contributed to a famine in 1961. This, in turn, prompted widespread rural–urban migration which eventually led to increased state controls on cityward migration. These problems resulted in the abandonment of the 'Great Leap Forward' in the early 1960s.

By the time of Mao's death in 1976 industrial growth rates were declining and agricultural production was not keeping pace with population growth. Peasants were increasingly allowed to produce and sell their own crops, rather than working within a communal system. The Chinese government also decided to move towards market mechanisms in some sectors by opening up parts of the economy to foreign investment. This was part of what the government termed 'market socialism'.

The first steps towards opening up were concentrated in Special Economic Zones (SEZs) located in the eastern coastal provinces of Guangdong and Fujian (Figure 3.3). Foreign investors were allowed

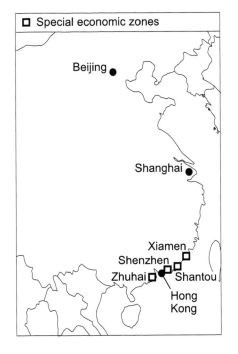

Figure 3.3 China's Special Economic Zones, 1979.

Source: adapted from Dicken (2007: Figure 7.5)

Map data © Maps in Minutes™ (1996)

to invest in these SEZs and the government offered tax concessions and preferencial land rents. The availability of a cheap and educated workforce was also attractive (Howell 1993). At first foreigners were not allowed to have complete ownership of these factories, so partnerships with Chinese state firms were set up. The economic success of these zones led to an expansion of the policy to other parts of the country, but the coastal zone remains the main focus of such activities, exacerbating existing regional inequalities between the coastal provinces, the central provinces and the western provinces (Wei and Ma 1996; Zhang and Han 2009) (see Chapter 4 for further discussion).

The political system still remains controlled by the Communist Party. Opposition voices are suppressed through controls on television, radio and newspapers. Policing of the Internet also takes place, with websites or chatrooms including anti-government sentiments being blocked or closed. Public displays of anti-government feeling are also restricted. The most well-known of these is the 1989 suppression

Plate 3.3 Pudong District, Shanghai.
Credit: Katie Willis

of the pro-democracy demonstrations in Tiananmen Square, Beijing. Over 1,000 people were killed during the violent police and army response, and many more were imprisoned.

Mao's rural development focus within a socialist framework has influenced government approaches to development in other parts of the world, for example Nyerere's policies in post-independence Tanzania (see below). In addition, some revolutionary groups in the South, such as Sendero Luminoso in Peru, have adopted Maoist principles (Alexander 1999).

African socialism or socialism in Africa

Ghana's independence from Great Britain in 1957 heralded a period of decolonization throughout the African continent. Many newly-independent nations drew on socialist ideas of development and progress in their focus on the role of the state in organizing and moulding economic activities. However, most did not fully embrace a socialist ideology in policy-making, not wanting to become too closely aligned with the Soviet Bloc countries of the USSR and Eastern Europe, but also not wanting to remain too politically-linked to previous Western colonial powers. Crawford Young (1982: 11) identifies three main ideologies that were adopted by the newly-independent African states: capitalist development (see Chapter 2), Marxist-Leninism (what Young refers to as 'Afro-Marxism') and 'populist socialism' (which is sometimes referred to as 'African socialism'). As with all such typologies, in some cases the classification process is very difficult. In this case, there are particular difficulties because of common disjunctures between the ideology a government espoused and the actual policies implemented.

Populist socialism was an attempt by a number of African nations to draw on the ideas of communal ownership of land and natural resources, and the traditional importance of collective working on the land. According to Africa's new leaders, such as Julius Nyerere in Tanzania and Leopold Sédar Senghor of Senegal, capitalism and class divisions were not part of African culture, but had been introduced by European colonialists. An 'African socialism' would, therefore, draw on the 'real' African roots, rather than following a Soviet-style model based on class struggle. The predominantly rural population of the African nations also led governments adopting this path to focus on rural development strategies, rather like Mao had done in China (Box 3.7). The key role of the state in directing development and in owning the means of production, such as factories and plantations, was similar to socialist models elsewhere, but African nations sought to adapt these models to African realities. Countries adopting these policies included Tanzania, Algeria, Ghana, Mali and Guinea.

In practice, the concept of African socialism came under increasing attacks in the late 1960s and early 1970s from both advocates of capitalist development, and those following a more Soviet style socialism (Young 1982: 100). Despite the claims to African 'authenticity', it was clear that ideals of a 'classless society' and the African communal way of life were not always as evident as had been claimed. These internal divisions, along with external pressures resulted in the gradual disintegration of the African populist socialist projects.

Afro-Marxism reflects an ideological approach drawing on the class analysis of Marx and focusing on the policies adopted in the USSR and associated states. This became more popular in the late 1960s and was a response to the perceived limitations of the African socialism model. Rather than trying to build development solutions based on the peculiarly African context, the argument was that the Soviet model of economic development and social welfare was applicable in all geographical locations. The 1959 Cuban Revolution was held up as an example of how countries in the South could successfully adopt a Soviet-style development path. Young (1982: 3) sums up the criticisms of African socialism as follows: 'The stress upon the communitarian heritage of African society was held to be naïve and the rejection of the class struggle mischievous and wrong.'

Marxist-Leninism was adopted in diverse ways among the African nations, but the importance of central state planning and state ownership of key units of production was at the heart of the policies. In reality, however, the state bureaucracy was too weak to implement planning systems following a Soviet model, and there were few companies which could be taken into state ownership. The role of the 'working classes' in the move to Marxist-Leninism was also limited because of the predominantly rural population (Young 1982). States adopting an Afro-Marxist approach included Congo-Brazzaville, Mozambique, Angola, Benin, Somalia and Ethiopia.

Box 3.7

Ujamaa *in Tanzania*

In 1961 Tanganyika became independent from Britain and in 1964 the Republic of Tanzania was created from the merging of Tanganyika and Zanzibar. The president of Tanzania, Julius Nyerere, wanted to promote development within Tanzania based on African traditions and values, rather than following a western model.

In 1967 with the Arusha Declaration the policy of *ujamaa* was declared. *Ujamaa* is a Swahili word meaning 'familyhood' and the basis of this development approach was rural development around communal farming in villages. As most of Tanzania's population was based in rural areas, Nyerere felt that the focus on rural development was appropriate. In addition, there was a tradition of communal activities, albeit on a much more limited scale than that envisaged under *ujamaa*.

Initially the policy encouraged peasants to move into *ujamaa* villages, or for those already resident in villages to increase communal, rather than private production. The government invested in water supplies, primary education and health provision in these villages. However, by 1970 progress was not regarded as sufficiently advanced, and the government made more interventions to make peasants move. As a result, the number of *ujamaa* villages increased from 1,956 in 1970 to 5,010 in 1974.

These shifts encouraged many White capitalist farmers to leave the country. There was also a policy of nationalization of the plantations producing crops such as sisal and coffee. While assistance was given to peasants in the form of fertilizer and some machinery, overall production remained rather low. Production targets were rarely met. Peasants spent the majority of their time working on their own private plots, rather than on communal village lands. This fitted within their own understandings of their lives and how they wanted to live. This was not, however, compatible with government plans for national development.

In 1973 a policy of compulsory villagization was announced. In some cases this involved extreme forms of coercion and violence. About 5 million Tanzanians were moved as part of this process which was seen as key to development. The *Daily News* newspaper reported:

> Mwalimu [Nyerere] has frequently reiterated during the last ten years the importance of people congregating in villages. Such proximity is a necessity of development. For when people abandon their isolation and come together in well-planned and laid out villages, they can be reached by social services, and they can effectively operate in co-operation. Only then can they begin to develop.
>
> (*Daily News*, 15 November 1974, in Hyden 1980: 131)

Because of production problems, the focus on communal village production was subsequently reduced and peasants were allowed to produce independently.

Source: adapted from Hyden (1980)

The collapse of the Soviet bloc in 1989 meant a severe cut in the assistance that could be provided to African socialist nations. However, even before this, socialist projects throughout the continent were experiencing problems, created both internally to the country, but also exacerbated by external forces. Limitations to economic growth were experienced as countries were unable to increase productivity through existing levels of education, skills and technology. In addition, opposition military activity (often supported by foreign governments) meant that stable conditions for economic growth were limited. For example, in Mozambique the opposition Renamo organization received funding from South Africa (Hanlon 1991). Increasing debt and economic problems forced many countries to accept restructuring packages associated with IMF loans and adopt neoliberal market-led policies (Sutton and Zaimeche 2002).

The rise of the Latin American left

Since the late 1990s, left-of-centre governments have been increasingly elected to power in many parts of Latin America. The election of Hugo Chávez in Venezuela in 1998 was followed by a wave of electoral success across the region, including Luiz Inácio Lula da Silva (often just known as Lula) in Brazil in 2002 and Evo Morales in Bolivia in 2005 (Plate 3.4). However, as Francisco Panizza (2009) outlines, this cannot be seen as a homogenous rise of socialist governments; rather there is a diversity of approaches. It has therefore been characterized as a 'pink tide' rather than a 'red tide', which socialism would imply.

The rise of the left in Latin America in the 2000s has been viewed as a response to the failures of the neoliberal model in the region. Periodic economic crises in the 1990s, for example in Mexico during 1994/5 and a rise in poverty in the period 1998–2002, created disillusionment among electorates about market-led development. Panizza (2009) also attributes the leftward trend to the opening up of democratic spaces in many countries, after decades of military rule. This provided opportunities for civil society organizations, such as the Brazilian Landless Rural Workers Movement (MST in Portuguese) to flourish. Such organizations tend to support left-leaning candidates.

Plate 3.4   Evo Morales and supporters, La Paz, Bolivia.

Source: © Boris Heger/Das Fotoarchiv/Specialist Stock

Ecuador, Bolivia and Venezuela have been most prominent in adopting what could be identified as socialist policies, including nationalization of productive resources and attempts to redistribute wealth (Box 3.8). However, there have also been attempts at regional cooperation through The Bolivarian Alliance for the Peoples of Our America (ALBA in its Spanish acronym). This is part of what Chávez has termed a 'Bolivarian Revolution', named after Simón Bolívar, a leader in the wars of independence against Spanish colonialism in the early nineteenth century. It has been set up as a counterpoint to free trade areas promoting neoliberal policies in the region.

Box 3.8

Evo Morales and the MAS government in Bolivia

Evo Morales was elected as president of Bolivia in 2005 and was re-elected in December 2009. Over 60 per cent of Bolivia's population belongs to an indigenous group, and Morales himself comes from an Aymara background. Prior to his election he was leader of the coca workers' union.

Morales' electoral success came after over two decades of market-led reforms in the country, including widespread privatization and increased foreign investment, particularly in infrastructure and oil and gas exploitation. This had led to widespread protests, including the so-called 'Water Wars' in Cochabamba, against the privatization of urban water supplies. There was also limited economic growth and the Bolivian economy was severely affected by regional economic crises such as that experienced by Argentina in 2001/2.

Morales' party is the Movement Towards Socialism (MAS in Spanish) and he has made his views about neoliberalism very apparent. For example, he is quoted as saying: 'Neoliberalism is the reproduction of savage and inhuman capitalism that continues to allow for the concentration of capital in few hands and that does not provide solutions for the majorities of the world' (quoted in and translated by Panizza, 2009: 183).

In power Morales and the MAS government have nationalized oil and gas reserves and have used the additional revenue for greater public expenditure in social programmes. There has also been legislation to promote the rights of indigenous peoples. However, attempts at constitutional change that seek to redistribute power have led to great opposition, particularly in the departments of Santa Cruz, Pando and Tarija, which have histories of greater foreign involvement and have larger non-indigenous populations.

Source: adapted from Panizza (2009); Reid (2007)

Summary

- Marxist definitions of 'development' are based on ideas of 'modernity'.
- Structuralist and dependency theories stress the importance of looking at the global economic system.
- Dependency theories argue that 'underdevelopment' is caused by unequal global power relations.
- State-socialist development models involve the primary role of the state as decision-maker.
- There have been attempts to introduce 'African socialism' with limited success.
- While state-led socialist models of development have been losing favour since the 1990s, in some countries, most notably in Latin America, state-led development is becoming more common.

Discussion questions

1   What were the main features of Marx's evolutionary approach to social development?

2   How do dependency theories challenge modernization approaches to development?

3   Why was import-substitution industrialization a key policy for structuralist theorists?

4   Why are state-led socialist development models less popular at the start of the twenty-first century compared to the post-Second World War period?

Further reading

Bradshaw, M. and A. Stenning (eds) (2004) *East Central Europe and the Former Soviet Union*, London: Pearson Prentice Hall. Excellent collection of chapters considering the transitional processes in Eastern Europe and the former Soviet Union.

Forbes, D. and N. Thrift (eds) (1987) *The Socialist Third World: Urban Development and Territorial Planning*, Oxford: Blackwell. Although this book is rather dated, it provides a very useful overview of the economic and regional development policies of a number of socialist regimes in the South. While most have now moved away from this form of development, there are many useful examples of how a socialist perspective has been put into development practice.

Frank, A.G. (1967) *Capitalism and Underdevelopment in Latin America*, London: Monthly Review Press. A classic presentation of the dependency approach.

Panizza, F. (2009) *Contemporary Latin America: Development and Democracy Beyond the Washington Consensus*, London: Zed Books. A clearly written account of the rise of 'the Latin American left' and its challenge to neoliberalism.

Useful websites

www.marxists.org   Marxists Internet Archive. Includes information about key Marxist writers and their publications. Includes sites on African socialism.

www.nacla.org   North American Congress on Latin America. A non-profit organization which provides information about Latin American social, political and economic issues.

www.socialistinternational.org   The Socialist International. Worldwide organization of socialist, social democratic and labour parties.

www.tol.cz   Transitions Online. News and information about the post-communist states of Eastern Europe and the former Soviet Union.

Grassroots development

- Basic needs
- Decentralization
- The role and increasing importance of NGOs
- Concepts of participation and empowerment
- Civil society and social capital

The theories discussed in the previous two chapters are based on activities at a national scale and involve decisions being made by governments without the direct involvement of the people. In this chapter the focus is far more on what have been termed 'bottom-up' or 'grassroots' approaches.

Basic needs

During the 1970s, the top-down approaches described in the previous chapters were increasingly recognized as having limited success in reducing the extent of poverty in large areas of the world. Rather than 'trickling down' to help the poorest, the benefits of economic development were being experienced largely by the richer countries and groups (Hettne 1995). This led to some rethinking of how development in terms of improving standards and quality of life was to be achieved.

Among large multilateral organizations, most notably the International Labour Organization (ILO) and the World Bank under the presidency of Robert McNamara, the concept of 'basic needs' was promoted. Under this approach, development policies were to be focused on the poorest people in society, rather than at a macro-level, which would indirectly help the poor. The approach became known as the Basic Needs Approach (BNA). At the ILO's World Employment Conference in 1976, the nature of 'basic needs' was outlined (see Box 4.1). These basic needs included not only the essentials for physical survival, but also access to services, employment and decision-making to provide a real basis for participation. While it was never acknowledged by the ILO, Leys (1996: 11–12) argues that the move towards basic needs was influenced by dependency theories (Chapter 3) in its premise that

Plate 4.1 Self-built house, informal settlement, Oaxaca City, Mexico.
Credit: Katie Willis

Box 4.1

ILO categories of basic needs

1 Basics of personal consumption – food, shelter, clothing;
2 access to essential services – clean water, sanitation, education, transport, healthcare;
3 access to paid employment;
4 qualitative needs – healthy and safe environment, ability to participate in decision-making.

Source: adapted from Hunt (1989: 265–6)

the perceived benefits of 'development' would not automatically trickle down to the poorest.

In policy terms, this approach advocated a focus on agricultural development and support for the urban informal sector, including greater research on labour-intensive production techniques that were appropriate for small-scale activities. In addition, a basic needs approach required public service provision to be expanded and developed to meet the needs of the poor. As Hunt (1989) stresses, most advocates of 'basic needs' were not calling for an end to the modernization project, rather that greater attention should be paid to smaller-scale activities and poorer sectors of society, alongside continued investment in large-scale infrastructure. Meeting the needs of the poor would not only help reduce poverty levels, but would also improve the education and skill levels of the population, with the concomitant potential for contributing to greater economic growth. In addition, as the poor get richer, their purchasing power rises, so benefiting domestic firms.

Plate 4.2   Peanut seller, Aleppo, Syria
Credit: Kelly Carmichael

Despite the widespread support for such an approach from many organizations and governments, there were a range of criticisms and implementation problems. Among these criticisms were the cost implications – improving public services is financially demanding, and in many cases governments were unwilling or unable to afford such expenditure. In addition, the 'basic needs' approach with its focus on small-scale agricultural production and informal sector activities was regarded by many as acting as a brake to rapid economic growth and continuing to trap Southern countries in primary production and low value-added manufacturing. As outlined in Chapter 1, there were also debates about how improvements in basic needs were going to be assessed (Hicks and Streeten 1979). John Friedmann (1992a: 59–66) outlines how the basic needs approach could have represented a real alternative to previous development approaches and policies, particularly with its focus on grassroots participation and wealth redistribution. However, he stresses that the application of the approach was generally very technocratic and focused on production and that by the 1980s the idea of 'basic needs' was losing its appeal.

Decentralization

The focus on the sub-national scale for development continued into the 1980s and 1990s. In market-led economies, there was a tendency to move away from central government activities and decision-making to a more decentralized approach. The rationale for adopting decentralization was both economic and political. In economic terms, it was argued that decentralizing government activities would lead to greater efficiency and cost-effectiveness. This clearly fits in with the neoliberal agenda. In political terms, by transferring decision-making to the more local level, people would be able to have a greater say in the decisions made about their services. This fitted in with the growth of 'good governance' as a key element of development policies and interventions (see Chapter 2).

This process of decentralization has been a major feature of policy-making in most countries, both North and South. Even in countries with an existing federal system where federal state governments have always had power to make significant decisions, such as Mexico, there has been a process of further decentralization. Within a neoliberal agenda, decentralization is regarded as a way of reducing state control, albeit in some cases just moving policy-making from central government to regional or local government.

As part of the shift from a state-oriented to market-oriented economy in China and transition economies (see Chapter 3), decentralization has been introduced for similar reasons. In China, following the revolution in 1949, the economy was divided into very small-scale production units from the village commune right up to the large state enterprises, but central government had the final power to make decisions. In terms of service provision, health and education expenditure, for example, was determined centrally and funds were provided to provincial and local governments. This meant that healthcare was accessible and affordable for the majority of the population. Approximately 85 per cent of China's villages in the late 1970s had health centres staffed by primary healthcare workers, known as 'barefoot doctors'. While patients had to pay for medical treatment, this was usually reimbursed. Complicated medical conditions were referred to health centres and hospitals in larger urban areas (Chetley 1995). As part of the reform process in China, provincial and local governments have been increasingly responsible for raising their own funds to provide services (Tang and Bloom 2000). This has had the effect of increasing regional inequalities in healthcare provision in China (Table 4.1). Similar patterns have been found elsewhere in the Global South where health sector reform has involved decentralization (Willis and Khan 2009).

Table 4.1 *Health reform and regional inequalities in China*

	Ratio of state health expenditure	
	East: Central	*East: West*
1986	1.44: 1	1.04: 1
1988	1.66: 1	1.17: 1
1990	1.71: 1	1.31: 1
1992	1.69: 1	1.16: 1
1994	1.98: 1	1.28: 1
1995	2.13: 1	1.40: 1
1996	2.09: 1	1.42: 1
1997	2.07: 1	1.55: 1

Source: based on unpublished figures from Center for Health Statistics Information, Ministry of Health, China

The involvement of local people in these forms of decentralized decision-making, while being very admirable in theoretical terms, does not always have the desired practical outcomes. As described below, participation, has become a key element of development theories and practice, but meaningful and widespread participation is much harder to achieve than was predicted.

NGOs as the development solution

The role of the central state as *the* key player in 'development' has always been debated, but it has become increasingly challenged from the 1980s onwards. This has been associated with the move towards neoliberal, market-led approaches as outlined in Chapter 2. As well as the devolution of elements of power and decision-making to the local state as described in the previous section, the move away from the top-down approaches has been associated in particular with the growth of non-governmental organizations (NGOs).

NGOs came to be seen as the panacea for 'development problems' by individuals from many different perspectives. First, however, it is important to realize the range of organizations which can fall under the NGO heading (Vakil 1997) (see Table 4.2). There are a number of axes along which NGOs can be divided. For example, NGOs can be very small-scale and operate in only one region or country. At the other extreme would be the very large organizations such as Oxfam and Save the Children Fund which are Northern-based, but have partner organizations throughout the world. Different NGOs will also have different forms of approach, either in terms of the overall development philosophy, or through the types of activities in which they get involved.

NGOs are often regarded as the answer to the perceived limitations of the state or the market in facilitating 'development' for a range of reasons (Lewis and Kanji 2009). First, it is argued that NGOs can provide services that are much more appropriate to local communities. This is because they work with populations at the grassroots to find out what facilities are required. In addition, they are able to provide such services more efficiently and effectively through drawing on local people's knowledge, and also using local materials. Because of the scale of operation and the linkages with local people, they are also able to react more quickly to local

Table 4.2 *Dimensions of NGO diversity*

Characteristic	Diversity
Location	North
	North in the South
	South
Level of operation	International
	Regional
	National
	Community
Orientation	Welfare activities and service provision
	Emergency relief
	Development education
	Participation and empowerment
	Self-sufficiency
	Advocacy
	Networking
Ownership	Non-membership support organizations
	Membership support organizations

Sources: adapted from Farrington and Bebbington (1993); Vakil (1997)

demands (Green and Matthias 1995). Finally, it is believed that NGOs are beneficial to non-material aspects of 'development', in particular processes of empowerment, participation and democratization. Because of the ways in which NGOs are embedded in local communities, it is argued that they have to be accountable to the local people. This means that local people have a greater say in what activities are carried out, and also that their participation in such activities creates an environment where empowerment is more likely. If such participation is expanded, these activities can help develop stronger civil society and contribute to processes of democratization (see below).

NGOs have, therefore, been interpreted as being the answer to all development issues, or what has been termed 'the magic bullet' (Edwards and Hulme 1995). Because of this enthusiasm, significant amounts of multilateral and bilateral aid are now channelled through NGOs as part of what has been termed the 'New Policy Agenda' (NPA). The NPA refers to the neoliberal approach within the international institutions, such as the World Bank. In 2008/9 £337 million of the UK government development aid (including debt relief) went to UK NGOs. This was from a total budget of nearly

Plate 4.3 Externally-funded irrigation scheme, Kenya.

Credit: Katie Willis

£7.2 billion. The British Red Cross received the largest amount of support (£32 million) with Voluntary Service Overseas (VSO) and Oxfam receiving £31 million and £25 million respectively (DFID 2009). Similarly, about 8 per cent of Australian ODA is channelled through NGOs, most notably large Australian NGOs such as Oxfam Australia and World Vision Australia (Govmonitor 2009).

The budgets of international NGOs (INGOs) such as those mentioned here, as well as philanthropic organizations such as The Bill and Melinda Gates Foundation, means that these non-governmental actors are becoming increasingly important players in the development aid field. For example, in 2005, World Vision International had an annual budget of over US$2 billion, which was more than the ODA budget of a number of Northern countries including Italy, Australia and Ireland (Koch 2008).

The number of NGOs throughout the world has increased very rapidly, partly because of the availability of funding, but also because of the lack of alternative support mechanisms for communities in need. This has been particularly the case as governments have reduced spending as part of their adoption of neoliberal policies, but may also be because of the weakness of state structures due to war or civil unrest. However, assessing the number of NGOs is very difficult. This is partly because of definitional difficulties, but it also reflects differing registration practices across the globe.

There are numerous examples of how NGOs have been able to provide services for communities when government assistance is not forthcoming or appropriate, and market-provided services are too expensive. Such NGO activities may include social welfare provision such as housing (Box 4.2), healthcare and education. In many cases it is clear that without NGO involvement many people's living conditions and quality of life would not be improved. However, the claims regarding NGOs have been revealed as being somewhat overstated. While many improvements can be made, NGOs alone cannot achieve everything that is expected of them.

Box 4.2

Viviendas del Hogar de Cristo Project, Guayaquil, Ecuador

Guayaquil (population about 2.5 million), like many Latin American cities, has experienced rapid growth due to rural–urban migration flows. As the government has been unable to provide sufficient housing for many residents, and individual households cannot afford to buy houses constructed by private sector companies, hundreds of thousands of people (about 60 per cent of the population) have resorted to building their own dwellings. These have often been of poor quality and have lacked access to basic services, such as water and sanitation.

In 1971 the NGO Viviendas del Hogar de Cristo (VHC) was set up by a Catholic priest to help address housing need in the city. The housing provided by VHC is largely bamboo and participants construct the houses themselves. The simple design of a house made of a wood frame with bamboo panels on a platform means that the houses can be constructed in a day. In addition, jobs are created in panel manufacturing plants and the bamboo is cultivated on sustainable plantations. In 1996 VHC won the World Habitat Awards. VHC now provides over 10,000 houses per year in Guayaquil and a number of other Ecuadorian cities.

continued

Financially, participants pay much less for these houses than equivalent state provision, and have access to credit through the NGO. Households are expected to pay back the cost of the house within two years. For those households unable to afford even low levels of repayment, the NGO has funds from donations allowing them to provide housing for free. About a third of houses are provided at no cost to the beneficiaries. Being of bamboo, the houses only last for about ten years, but they act as an excellent form of shelter in the short term.

Services, such as water and electricity are not provided as part of the scheme; residents must work together to lobby local authorities to supply basic urban services. Since 2001, VHC has expanded its operations to include microcredit, health and education services. Some of these have been in collaboration with the Ecuadorian government. For example, in 2007, VHC worked with the Ministry of Economic and Social Inclusion to build and run a refuge for women and children fleeing domestic violence. VHC also works in Chile.

Sources: adapted from Diacon (1998); Hogar de Cristo (2010)

Empowerment

NGOs' ability to 'empower' individuals and communities has been an important part of the enthusiasm with which NGOs have been greeted (Kilby 2006). As Rowlands (1997, 1998) has highlighted, 'empowerment' has become one of the key buzzwords in development policy since the early 1990s, but it is a term with diverse and contested meanings. At the heart of the concept is the idea of having greater power and therefore more control over your own life, but as Rowlands stresses, this does not recognize the different ways in which 'power' can be defined.

The kind of power that we often think about is the power to be able to get other people to do what we want, or the power that other people have to make us do something. This can be termed 'power over' and is often regarded as the most important form of power because it is associated with processes of marginalization and exclusion through which groups are portrayed as 'powerless'. However, there are other dimensions of power that can be identified and which should be considered as part of the development process. Rowlands terms these 'power to', 'power with' and 'power within' (see Box 4.3). All of these forms of power are linked, but a recognition of the diversity of power

---

Box 4.3

Dimensions of power

Power over   The ability to dominate. This form of power is finite, so that if someone obtains more power then it automatically leads to someone else having less power.

Power to   The ability to see possibilities for change.

Power with   The power that comes from individuals working together collectively to achieve common goals.

Power within   Feelings of self-worth and self-esteem that come from within individuals.

Source: adapted from Rowlands (1997, 1998)

---

beyond 'power over', helps in the construction of policies and programmes to assist the 'powerless'.

A key element of 'empowerment' as a development outcome is what forms of intervention can lead to 'empowerment'. It is often claimed that NGOs can 'empower' communities, but in reality this is not the case. This is because empowerment is something that comes from within (Townsend *et al.* 1999). While NGOs may be able to provide a context within which a process of empowerment is possible, it is only individuals who can choose to take those opportunities and to use them. For example, illiteracy is often regarded as an obstacle to participation in waged work and political life. NGOs may be able to provide facilities and teachers to help individuals develop their literacy skills, but individuals themselves have to want to participate and to use their newly-acquired skills. This does not mean that disadvantage and exclusion are the fault of individuals, there are clearly structural constraints, but it does mean that NGOs cannot be viewed as direct channels for empowerment; rather they can help set up conditions within which individuals and groups can empower themselves.

Participation

One of the key routes through which empowerment is meant to be achieved is through 'participation'. Grassroots development is often termed 'participatory', but it is important to recognize what this means. Participation is usually used as an umbrella term to refer to the involvement of local people in development activities, often NGO based. However, this participation varies in nature (Table 4.3).

Participation can take place in a number of ways and at different stages in the setting up of development projects. With a growing awareness of the power relations involved in development research, NGO practitioners and researchers have begun to use a range of different methods in order to find out more about the understandings local people have of a range of topics. These approaches have often been termed 'participatory'. Participatory rural appraisal (PRA) is an approach which many organizations have now adopted (Chambers 1997). The basic premise of PRA is that villagers in the Global South have a knowledge and awareness of their environment which should be taken into account, rather than using 'outsider' forms of knowledge and understandings. For example, when trying to assess economic status within a village, a purely income-based approach would ask questions about earnings from wages and market activities. In contrast, PRA approaches to wealth ranking would ask villagers to come up with their own indicators of wealth in the area and then use these to rank household wealth (Hargreaves *et al.* 2007). The methods used are therefore appropriate to the context. PRA techniques have been increasingly adopted in urban settings, using the same principles. This has been termed PUA (Participatory Urban Appraisal) (Moser and McIlwaine 1999).

Participation can also be used to refer to the involvement of local people in the actual agenda setting of development organizations. To be fully participatory, the agenda needs to be set by the communities

Table 4.3 *Dimensions of participation*

Dimension	Meaning
Appraisal	Way of understanding the local community and their understandings of wider processes, e.g. PRA, PUA
Agenda setting	Involvement of local community in decisions about development policies; consulted and listened to from the start, not brought in once policy has been decided upon
Efficiency	Involvement of local community in projects, e.g. building schools
Empowerment	Participation leads to greater self-awareness and confidence; contributions to development of democracy

involved, rather than outside agencies deciding on the priorities to be addressed and then working with local people to achieve them.

The two previous interpretations of participation can be regarded as contributing to processes which address power inequalities in development and also attempt to focus more on grassroots forms of knowledge than external, particularly Northern, ones. The focus on participation as a route to empowerment is linked to this. As NGOs have been placed at the foreground of such activities, the concept of participation has been promoted as a way of giving local people greater decision-making power and influence (Box 4.4). However, far too often this form of participation is not achieved; instead local people are involved in meetings or contributing labour, but this is not participation in the wider sense which can be linked to empowerment.

Bill Cooke and Uma Kothari (2001) refer to participation being the 'new tyranny' in development work. They argue that the notion of participation is included in every dimension of development policy, but there is no recognition of:

- the time and energy requirements for local people to participate;
- the heterogeneity of local populations meaning that 'community participation' does not always involve all sectors of population;
- just being involved does not necessarily lead to 'empowerment';
- focusing at a micro level can often lead to a failure to recognize much wider structures of disadvantage and oppression.

The fact that 'participation' and 'participatory approaches' are encouraged by multilateral organizations such as the World Bank (Francis 2001) suggests that these are ideas which have been taken on board, but the dimensions of participation that could challenge existing practices and power relations are not engaged with. When individual projects are examined, the limitations of the participation discourse become apparent (see Box 4.5). David Mosse in his critique of current participation policies, states that 'participatory approaches have proved compatible with top-down planning systems, and have not heralded changes in prevailing institutional practices of development' (2001: 17). However, contributors to the volume edited by Sam Hickey and Gile Mohan (2004) argue that participation can still be transformative within certain settings.

Some of the problems in implementing truly participatory development projects come from the pressures of the so-called

Box 4.4

Barefoot College, India

Barefoot College is a rural development organization based in the Ajmer District of Rajasthan in north-west India. It was started in 1972 by urban-based development practitioners who set up the Social Work and Research Centre. The name was changed later and it is now run by local people.

A key philosophy of the organization is providing appropriate training and technology to be used in the surrounding rural areas. Basic literacy and health skills are taught, but individuals can then go on to train in fields such as computing, teaching and healthcare, or as engineers skilled in working with appropriate technology. This includes solar power and hand pumps. By using this low-cost technology, which is easy to maintain and use, Barefoot College has been able to provide services to about 100 villages and over 100,000 people. In addition to the main campus the College works through 8 field centres. Once trained, individuals work as 'Barefoot professionals'. Providing such training helps to reduce outmigration, as well as providing services that are required locally.

The College receives funding from a range of Indian national and state government agencies, as well as from United Nations agencies and a number of Northern governments and Northern NGOs. Income is also received by charging for the services it provides. The College charges a nominal fee for its services to promote a 'sense of ownership' among participants. For example, to benefit from solar power projects, households must make a contribution, but the level of that contribution is set based on their current expenditure on candles, kerosene and torch batteries.

The locally-based and participatory approach to rural development is also reflected in the College policy towards accountability and transparency. To assist in this process, the College has, since 1997, held 'transparency *melas*'. These are public meetings where the financial records of the organization are presented and villagers can ask questions about the use of funds. The horizontal structure of the organization is also reflected in the wages. Nobody earns more than US$150 per month and the ratio of the lowest wage to the highest wage is no more than 1:2.

Source: adapted from Barefoot College (2010)

Box 4.5

NGOs and participation in Tanzania

Under the one-party system in Tanzania in the 1960s and 1970s (see Chapter 3) there were very few opportunities for organizing outside government-linked groups. Trade unions, cooperatives, women's groups and other autonomous organizations were banned and it was only religious groups (both Christian and Muslim) that were able to operate, albeit within certain restrictive boundaries.

Following the move to a multi-party system and the adoption of structural adjustment policies (see Chapter 2) in the 1980s, the number of NGOs in Tanzania increased rapidly. In 1986–90 there were 25 NGOs registered with the Ministry of Home Affairs, but by 1995 this number had increased to 604. Of course, the actual number of NGOs is likely to be much higher as not all groups will register, but the quantitative shift is very apparent.

While the government and the NGO leaders often talk about empowerment and participation in relation to NGOs, the activities of the NGOs often fall short of these goals. Most NGOs focus on service provision such as healthcare or education, or improvements in income-generating capacity. In the context of structural adjustment and the decline in state-provided services, this focus is understandable and important, but it does not necessarily deal with the themes of participation and empowerment.

The geographical distribution of NGO activity in Tanzania reflects existing distributions of wealth and influence. NGOs are often presented as meeting the needs of the very poorest in society, but in Tanzania in 1994, nearly 60 per cent of the registered NGOs were located in Dar es Salaam which is the richest part of the country. Another 10 per cent were based in Arusha and 6 per cent in the Kilimanjaro region, both among the wealthier regions. The poorer regions were very poorly served by registered NGOs, leaving them less able to access funds and support.

At a very local level, this focus on the richer elements of society continues. In Hai District on the western slopes of Mount Kilimanjaro, the vast majority (79 per cent) of the 97 groups operating in 1996–7 focused on women's income-generating schemes. Such schemes can help marginalized groups to improve their incomes and therefore the health and education of household members. In some cases, feelings of empowerment may also be an outcome. However, women's participation in these groups was often limited by time or financial constraints. NGO membership was dominated by women from middle-income groups who were older, married and Christian rather than Muslim.

continued

The following quotations provide some indication of why poorer women were not able to participate in NGO activities:

CM: Why didn't your wife join the Kalili women's group?

Respondent: We knew about it but we couldn't afford it. We knew how much it would cost.

Respondent 2: I'm a member of Nkwarungo church, but I'm not a member of the women's department. I would very much like to join, but the problem is time. I have to stay here and look after the children, the livestock, go to the market – I don't have time to attend meetings at the church.

Source: adapted from Mercer (1999)

'development industry'. The need for rapid, quantifiable results, encourages project managers to focus on certain forms of project with tangible outcomes, rather than addressing deep-rooted inequalities which cannot be easily measured. As Edwards and Hulme (1995: 9) argue, NGOs have multiple accountabilities; '"downwards" to their partners, beneficiaries, staff and supporters; and "upwards" to their trustees, donors and host governments'. Managing these tensions is a difficult endeavour and one in which the long-term participatory strategy of projects is likely to suffer. Dependence on external aid assistance (Tables 4.4 and 4.5) means that many projects are more likely to react to the requirements and preferred activities of the potential donors (NGOs, foreign governments and multilateral organizations) than local people (Ahmad 2006; Hashemi 1995; Hulme and Edwards 1997; Porter 2003).

This donor reliance threatens the 'non-governmental' nature of NGOs as their capacity for autonomous action is reduced. This can be seen throughout the world. For example, in the Global North some NGOs are increasingly acting as service providers for local and national governments. While this can have very beneficial effects in terms of the quality of the service provided and the involvement of local people, it can result in NGO activities being framed primarily by the state funders, rather than by the supposed beneficiaries on the ground.

The undermining of supposed NGO neutrality is particularly evident in conflict zones (Duffield 2007). Róisín Shannon (2009) examines

Table 4.4 *Dependence on external aid, 1990 and 2005*

	Official Development Assistance (ODA) received[a]		As % of GDP	
	Total (US$m) 2005	*Per capita (US$) 2005*	*1990*	*2005*
Developing Countries	86,043.0	16.5	1.4	0.9
Least Developed Countries	25,979.5	33.9	11.8	9.3
Arab States	29,612.0	94.3	2.9	3.0
East Asia & Pacific	9,541.6	4.9	0.8	0.2
Latin America & Caribbean	6,249.5	11.3	0.5	0.3
South Asia	9,937.5	6.3	1.2	0.8
Sub-Saharan Africa	30,167.7	41.7	5.7	5.1
Central & Eastern Europe and CIS	5,299.4	13.1	–	0.3
High income	–	–	–	–
Middle income	42,242.2	13.7	0.7	1.3
Low income	44,123.0	18.2	4.1	3.2
World	**106,372.9**	**16.3**	**0.3**	**0.2**

Source: adapted from UNDP (2007: Table 18, pp. 290–3)

Note

[a] Net disbursements.

Table 4.5 *Countries most reliant on official development assistance, 2005*

	Official development assistance received[a] *as % GDP*
Solomon Islands	66.5
Timor-Leste	52.9
São Tomé and Principe	45.2
Burundi	45.6
Eritrea	36.6
Sierra Leone	28.8
Congo	28.5
Malawi	27.8
Occupied Palestinian Territories	27.4
Rwanda	26.7

Source: adapted from UNDP (2007: Table 18, pp. 290–3)

Note

[a] Net disbursements.

the role of NGOs providing humanitarian and development assistance in Afghanistan following the US-led invasion in 2001. She concludes that NGOs, by accepting funding from foreign governments involved in the invasion and working alongside the Afghan government, have been seen to take sides. In 2009, about US$1.5 billion of USAID's funds went to Afghanistan. While a significant proportion went directly to the Afghan government, NGOs were also major recipients (USAID 2010). The perceived impartiality of some NGOs has an impact on their ability to work with vulnerable populations in many parts of the country (Shannon 2009).

Civil society

As well as providing grassroots level possibilities for involvement and participation, NGOs can, it has been argued, contribute to wider processes of democratization through their contribution to civil society (Lewis and Kanji 2009). The concept of civil society has received increasing attention in recent decades, particularly in relation to the move from authoritarian forms of government, such as military regimes in South America or communist governments in the 'Second World' to multi-party democracies. However, the concept is also applied to other forms of national politics and the concept of 'global civil society' has also gained popularity (see Chapter 7).

The idea of 'civil society' is certainly not new, having its roots in seventeenth- and eighteenth-century social and political philosophy. However, during the 1990s it became a 'buzzword' in development circles and the strengthening of 'civil society' has become a key element of many development approaches (McIlwaine 1998; Van Rooy 1998, 2008). The 2002 *Human Development Report* had as its theme 'Deepening democracy in a fragmented world' and stressed the importance of freedom and choice in development:

> Politics matter for human development because people everywhere want to be free to determine their destinies, express their views and participate in the decisions that shape their lives. These capabilities are just as important for human development – for expanding people's choices – as being able to read or enjoy good health.
>
> (UNDP 2002: 1)

One of the indicators that UNDP uses to measure this level of freedom is the strength of civil society.

As with many concepts gaining great popularity, there are a wide variety of definitions of 'civil society'. Broadly it covers activities and organizations that are separate from the state and from the market (Box 4.6), but the debates rage about the precise boundaries. This diversity of interpretation contributes to its widespread appeal (Van Rooy 2008). However, as Cathy McIlwaine states in her overview of civil society and development geography:

> While there is little precise agreement as to which activities should be included within civil society, particularly in terms of political and economic associations . . . they generally refer to voluntary organizations, community groups, trade unions, church groups, co-operatives, business, professional, and philanthropic organizations, and a range of other NGOs . . . . Although civil society is usually defined as made up of these various groups, there has also been a tendency to view NGOs as primary 'vehicles' or 'agents' of civil society.
>
> (McIlwaine 1998: 416)

This focus on NGOs within civil society is evident from the UNDP measures. One of the 'objective indicators of governance' used by the UNDP is the number of NGOs (UNDP 2002: 442–5).

Plate 4.4 Anti-logging demonstration, Ormoc City, Leyte island, Philippines
Credit: © Nigel Dickinson/Specialist Stock

Box 4.6

Living Wage campaigns

Campaigns to introduce 'living wages' are important examples of the role of civil society groups in seeking to improve the lives of marginalized and excluded populations. While the idea of a living wage dates back to 1870s industrial Britain, the recent focus is usually attributed to the actions of unions and faith-based groups in Baltimore, USA.

During the 1980s, Baltimore's city government sought to promote regeneration of the city by encouraging investment and redevelopment on the Baltimore waterfront. However, while the scheme can be seen as a success in some senses, the jobs that were created tended to be low-paid, part-time and insecure jobs in the service sector, particularly in hotels and retailing. This shift in employment is common throughout the world in locations that have shifted from industry to services. A coalition between unions and church groups, particularly those with a large African-American congregation, lobbied large employers in the Inner Harbor development to pay a living wage to its employees. When this proved to be unsuccessful, the coalition turned its attention to the government, arguing that large businesses in the development had benefited from public subsidies, so the government had a responsibility to ensure that workers in these businesses were not being exploited. In 1994 the city agreed to a living wage ordinance, which required companies contracted by the city government to pay a living wage. In July 1995 this was US$6.10 per hour compared with the federal minimum wage at that time of US$4.25. Since then, living wage laws have been implemented in about 130 US cities.

The Baltimore case has inspired movements elsewhere in the world. For example, London Citizens is an umbrella organization of over 100 civil society organizations. One of its key campaigns is for a London Living Wage, which in 2009 was calculated at £7.60 per hour, compared with the National Minimum Wage of £5.80. Through lobbying politicians and company boards, as well as staging public protests, the London Living Wage campaign has achieved notable successes. They provide training in campaigning skills for local community organizers. The concept of a 'living wage' is also used by workers in the Global South, such as textile workers in South Asia.

Source: adapted from Fairris and Reich (2005); Labour Behind the Label (2010); London Citizens (2010); Luce (2005); Pattison (2008); Welsh (2000); Wills (2009, 2010)

Claire Mercer (2002) argues that there are three main reasons that NGOs are interpreted as key in promoting democracy:

- They provide an opportunity for people not involved in state organizations to 'have a say'. These voices may provide a challenge to the state.
- They supposedly work with the poor and marginalized (although, see pp. 108–20).
- They can provide a check to state power by means of dissent.

However, as Mercer argues, far too often NGOs are assumed to provide these forms of alternative political activity, but in reality this is not achieved. She states that just counting the increase in NGOs does not provide us with a satisfactory measure of expanding civil society. As outlined above, NGOs are not always able to maintain their automony or independence in relation to governments.

Social capital

The increasing focus on civil society and particular NGOs in development policy has been strongly associated with the focus on ideas of 'social capital' as a key asset for individuals and communities in the 'development' process (Bebbington 2008; Fine 2010). As with 'civil society', 'social capital' is a highly-contested concept, but at its roots is the idea of social relations between individuals and groups. These relations are based on trust and there are expectations regarding how you should behave in these social interactions.

Social capital is regarded as another asset which can be used by individuals or groups to contribute to their economic and social advancement, similar to economic capital (such as money and property) or human capital (education and health). Thus, individuals without sufficient or suitable social capital can be marginalized or vulnerable. Social capital is usually identified as being either 'bonding capital', describing links between individuals and groups of similar backgrounds or from the local community, or 'bridging capital', which is links outside the immediate group or locality (Pelling 2002). 'Linking capital' is sometimes used to refer to connections to individuals in positions of authority, such as local government oficials, politicians or financial institution representatives (Grootaert et al. 2004).

While the concept of links between people that can be mobilized to meet certain individual or group needs is certainly not restricted to a particular development approach, the term 'social capital' has most clearly been adopted as part of the neoliberal agenda of the international financial institutions (IFIs). For example, the World Bank began using the term widely during the 1990s and declared that 'social capital' was the 'missing link' in development (Grootaert 1998), meaning that it was the final piece in the development jigsaw; the generation of social capital along with improvements in social well-being and economic advancement would be a mutually reinforcing process. The World Bank's definition of 'social capital' is 'the institutions, relationships, and norms that shape the quality and quantity of a society's social interactions . . . . Social capital is not just the sum of the institutions which underpin a society – it is the glue that holds them together' (World Bank 2010d).

Networks of friends, kin and acquaintances are clearly important in helping individuals throughout the world meet their basic needs, as well as forming the basis for community organizations that make up civil society (Bebbington 1997). Social capital can also be important in communities' responses to emergencies (see Box 4.6). However, the way in which the concept has been used by IFIs and many governments, for example, has stressed the role of the individual and the local community in developing these networks, and has left out considerations of larger structural issues about access to finance and opportunities (Bebbington 2007). In the case of Santo Domingo (Box 4.6), it is clear that the community is reasonably served by social capital, but the problem is how to help this community in the longer term. For this, there need to be greater links outside the community to people and organizations (including government) that have the financial and technical resources to help. Processes of exclusion on the basis of ethnicity, gender and class (among other things) also need to be addressed (see Chapter 5).

Despite the continued widespread use of the term 'social capital', many organizations, including the World Bank, no longer focus on it within their policy-making and interpretations of development issues. Ben Fine (2008, 2010), who has been a long-standing critic of social capital, argues that this reflects the concept's lack of utility as a tool for both theory and policy.

Box 4.7

Social capital and hazard vulnerability in Santo Domingo

The urban district of Los Manguitos in Santo Domingo, Dominican Republic is vulnerable to natural hazards because of its location on a river and also because of the poverty of its residents. Over 40 per cent of the population are classified as 'poor' by the Dominican government. Los Manguitos began as a squatter settlement in 1975 and by 1999 there were over 33,000 residents.

In 1998 Hurricane Georges hit the area, and widespread damage was caused by the resultant flooding and high winds. The population of Los Manguitos was able to cope with many of the resulting problems by calling on neighbours to help out and also by using local organizations. Householders who were forced to leave their homes often moved in with neighbours or slept in the local church which was set up as a shelter. In addition, local community groups organized the evacuation process and helped clean up the debris and carry out repairs afterwards. This was without the assistance of the government or international aid.

This mobilization of social capital at times of emergency is an important indication of how these ties can help poor vulnerable communities. However, links to external organizations, either national or international, would help reduce vulnerability to hazards and would also help this community if disaster struck again. At the moment, the bonding capital between individuals in the community is strong and there is evidence of bridging capital through the creation of community organizations. There needs to be more bridging capital, allowing local residents and groups to access money and assistance from outside Los Manguitos.

Source: adapted from Pelling (2002)

Grassroots organizations and post-development

In this chapter we have seen how a focus on the local-level or small-scale has been advocated at different times and for different reasons in the search for 'development'. For 'post-development' theorists (see Chapter 1) this scale is also important. Rather than imposing ideas of 'progress' and 'development' on individuals and communities throughout the world, people themselves should be able to choose the way they want to live without being made to feel that they are somehow 'inferior' or 'backward' by not following a pattern that has been adopted elsewhere.

Concepts of grassroots activities and participation have become very much part of the lexicon of development, even within large-scale multilateral organizations. However, as was argued above, these shifts in scale remain firmly focused on Northern-based neoliberal concepts of 'development' and 'progress'. Richard Peet and Elaine Hartwick (2009: 217) summarize the arguments of Majid Rahnema in relation to this process:

> Rahnema noted that governments and development institutions became interested . . . because participation was no longer perceived as a threat, was politically and economically attractive, was a good fundraising device, and was part of a move toward the privatization of development as part of neoliberalism.

For post-development theorists and practitioners like Rahnema (who has been a UN Representative in Mali), the key is to make the act of participation actually mean something. Rather than incorporating individual and community views into programmes and projects just 'for appearance', local views should be prioritized in the development of any policies.

In addition, rather than attempting to consider and act at a 'global scale', the focus should be on local views and actions (Esteva and Prakash 1997). This does not rule out alliances and transnational networks (see Chapter 7). In their call to 'think and act locally', Gustavo Esteva and Madhu Suri Prakash (1997: 282) argue:

> [W]hat is needed is exactly the opposite [of acting globally]: people thinking and acting locally, while forging solidarity with other forces that share this *opposition* to the 'global thinking' and 'global forces' threatening local spaces. For its strength, the struggle against Goliath enemies demands that there be no deviation from local inspirations and firmly rooted local thought. When local movements or initiatives lose the ground under their feet, moving their struggle into the enemy's territory – global arena constructed by global thinking – they become minor players in the global game, doomed to lose their battles.
>
> (Emphasis in the original)

The concept of 'post-development' has been hotly debated, with many development theorists challenging the way in which authors such as Arturo Escobar and Gustavo Esteva have talked about 'development'. Jonathan Rigg (2003) in his study of development in Southeast Asia, highlights how post-developmentalists tend to talk about 'development' as if it is only Eurocentric modernization style development. While it is clear that in many policy arenas the importance of real grassroots activity and participation is not really grasped, it is a vast generalization to say that *all* development fits into this category (see also Corbridge 1998; Nederveen Pieterse 2000). The post-development approach will be discussed further in Chapter 8.

Summary

- From the 1970s onwards growing attention was paid to bottom-up or grassroots development.
- Decentralization and devolution are often claimed to be more efficient than larger-scale activities, so are part of a neoliberal agenda.
- Non-governmental organizations have become key actors in the grassroots approaches.
- NGOs are often thought to be highly participatory and contribute to empowerment, but this is not always the case because of existing social relations and the limits on NGO activity.
- Civil society groups like NGOs are regarded as being important in the move towards greater democracy throughout the world.

Discussion questions

1   Why have NGOs been regarded as the solution to development problems?

2   How do grassroots approaches and neoliberalism fit together?

3   What limits are there on the effectiveness of NGOs?

4   What is social capital and how can it contribute to local development?

Further reading

Bebbington, A.J., S. Hickey and D.C. Mitlin (eds) (2008) *Can NGOs Make a Difference? The Challenge of Development Alternatives*, London: Zed Books. Useful collection of chapters considering how far NGOs can provide alternatives to mainstream neoliberal development models.

Chambers, R. (1997) *Whose Reality Counts? Putting the First Last*, London: Intermediate Technology Books. A very accessible book dealing with the importance of indigenous knowledge and how development 'experts' should listen to the people with whom they are working.

Cooke, B. and Kothari, U. (eds) (2001) *Participation: The New Tyranny?*, London: Zed Books. Excellent collection of chapters on what participation means and how it has been adopted as a standard aim of many development projects.

Hickey, S. and G. Mohan (eds) (2004) *Participation: From Tyranny to Transformation*, London: Zed Books. A response to the Cooke and Kothari collection that seeks to examine the potential for development policies to fulfil their participatory ambitions.

Lewis, D. and N. Kanji (2009) *Non-Governmental Organizations and Development*, London: Routledge. A clear introduction to the changing role of NGOs in development theory and practice.

Useful websites

www.ausaid.gov.au   AusAID website. Includes information about the AusAID policies towards NGOs.

www.eldis.org/go/topics/resource-guides/participation   Collection of up-to-date resources on participation run by the Institute of Development Studies, University of Sussex.

www.oecd.org   OECD. Provides up-to-date information about aid flows.

www.oneworld.net   One World website. An excellent site for up-to-date development information. Following the 'Partners' link will provide access to information about over 1,600 partner organizations.

5 Social and cultural dimensions of development

- Development as social evolution
- Weber, rationality and the Protestant ethic
- Ethnodevelopment
- Gender and development
- Rights-based development

The previous chapters, especially Chapters 2 and 3, have focused more directly on economic aspects of 'development'. However, it is crucial to recognize social and cultural elements. Of course, as discussed in Chapter 1, particularly in relation to postcolonialism, the economic-focused discussions of the previous chapters are framed by the social and cultural norms of the people and institutions involved. However, in this chapter we pay explicit attention to social and cultural dimensions of development. This is not just because social and cultural variables affect economic growth, but also because social and cultural norms and expectations need to be considered in their own right (Radcliffe 2006; Schech and Haggis 2000).

Social evolution

'Modernity', as a development goal, has more than just economic elements. As Rostow discussed in relation to the linear model of economic growth, switches from one stage to another may be assessed through economic output, but many of the driving forces of a shift from subsistence agricultural societies to urban industrial ones relate mainly to social structures. As Worsley (1999) points out, this element of Rostow's thinking is often excluded from the characterizations of his work. In Rostow's argument, for take-off to be achieved, there need to be shifts in the nature of national elites. Rather than elite groups using their money for family and clan purposes, for industrial development to take place there needs to be a move to investment in new infrastructure and means of production (see Chapter 2 for more details). Thus, modernization theories refer not only to economic changes, but to social transformations as well.

The social evolutionary ideas which form part of Rostow's model have a basis in the nineteenth-century theories about social change based on the capitalist experience in Western Europe. During this period, many social theorists, in particular Herbert Spencer (1975 [1876–9]), drew on Charles Darwin's Theory of Evolution to examine and explain the shifting patterns of social organization. Just as biological organisms become increasingly complex through processes of competition and natural selection, so, such theorists suggested, did societies. This approach, which was termed 'naturalism' conceived of societies as being like organisms. Thus, societies which were best able to adapt to environmental and other conditions, became 'winners', and those which did not were doomed to remain 'traditional' (see discussion of environmental determinism in Chapter 6). Such theories were used to justify colonial expansion through which dominant groups were able to impose their will on those societies which had been 'less successful' (see Chapter 1). In addition, '[t]he Darwinian notion of survival of the fittest, applied to human societies, was used to legitimate laissez-faire, market systems, the private ownership of productive resources, and social inequality (Peet and Hartwick 2009: 106).

Emile Durkheim, a French sociologist (1858–1917), looked at the transition from 'traditional' to 'modern' societies in Western Europe, and while he adopted some forms of naturalism in his approach, he did not imply that humans were unable to exercise agency. In his view, societies were constructed of a set of moral and ethical norms into which individuals were born. Societies would have different forms of punishment or control to ensure that the society remained in equilibrium. In this sense, therefore, Durkheim adopted the ideas of natural systems in assuming that harmony and equilibrium were constantly striven for.

In his 1893 book, *The Division of Labour in Society*, Durkheim argued that within 'traditional societies' individuals were part of tight-knit communities, often consisting of kin and clan. Attempts to break from this mould were punished through severe forms of retribution. In contrast, in 'modern societies', there was much greater individualism because of the need for a division of labour. As societies become more complex it is impossible for one person to do everything to ensure daily survival and the creation of profit, so people take on particular roles within the economy allowing for expansion. Despite the individualism in this sense, society's members are, according to Durkheim, still part of a highly-integrated whole. However, Durkheim did recognize that this form of integration was not always present; where individuals became disengaged from one another, particularly when change was too quick or not regulated, a state of 'anomie' was created. 'Anomie' was understood by Durkheim to be,

> a feeling of rootlessness and aimlessness which, furthermore, was characterised by a lack of moral guidelines. The breakdown of the traditional orders, which were supported by religion, would result in many people feeling that their lives had lost meaning; they would feel isolated without guidelines for moral behaviour.
>
> (Martinussen 1997: 26)

This was regarded as detrimental to the continued functioning of the society, so various institutions should be set up to encourage social interaction. In certain situations anomie could be associated with conflict.

The concept of a harmonious society acting as an organism was also used by Talcott Parsons (1902–79), an American sociologist who built on the work of Durkheim in his analysis of social development (1951). His understandings of society were very technocratic, in that he saw the harmony and continued flourishing of society to be ensured through appropriate institutions. He argued (1966) that over time, societies adapted and evolved to become complex. These changes could be triggered either by external influence introducing new technologies or cultural forms, or internally. He viewed societies, as Durkheim did, as moving from 'traditional' to 'modern' and identified a number of 'pattern variables' to distinguish between the two. For example, in 'traditional' societies status was gained through kinship, ethnicity or gender, i.e. it was ascribed. In contrast, in 'modern' societies, individuals earned status through what they did, such as gaining formal qualifications or different forms of paid work (Brohman 1996: 19–21).

Parsons' structural functionalist approach to explaining development has been criticized from a number of perspectives (see Preston 1996; Peet and Hartwick 2009 for more detailed discussions). First, the approach is based on the idea that a society's codes are 'pre-given' and that individuals just have to fit into these structures. There is no recognition of how these codes and norms are, in fact, socially constructed and reflect power relations. In addition, the approach is a very conservative one, stressing the need for harmony and dismissing conflict as undesirable. In some senses, however, conflict could be a key factor in creating and conditions for more radical change which could bring greater social benefits to a larger number of people. For example, independence struggles throughout the world were based on challenging existing structures and ways of doing things. Finally, and linked to the previous point, the approach is Eurocentric and sets up the 'Western experience' as the model that all other parts of the world should follow.

Weber, rationalism and the Protestant ethic

Explanations for differences in 'development' that are based at the level of societies have been criticized for having little or no recognition of agency (see Chapter 1) on the part of individuals and groups making up these societies. Instead, humans are viewed as other animals, acting purely on instinct (in the case of naturalism), or subsumed within the operation of wider society (as in structural functionalism). The nineteenth-century German theorist Max Weber approached social change from a different perspective. In contrast to Durkheim's focus on society as an organic whole, Weber used the individual as the starting point (Preston 1996).

Based on his analysis of German society in the late nineteenth century, he argued that over time, different elements of social, political and economic life would be brought under the control of rational thought (Roxborough 1979). This means that rather than relying on 'superstition' or 'tradition' in making decisions, individuals would base their choices on logical analyses of the situation. This does not mean that Weber saw no role for religion in social change and the development of capitalism. In his book *The Protestant Ethic and the Spirit of Capitalism* Weber described how Calvinist Protestantism was associated with the growth of industry in Germany. Within Calvinism, there is a focus on the relationship of the individual with God and an emphasis on delayed gratification. According to Weber, this individualism and the work ethic associated with the requirement to delay personal rewards led to rapid economic growth.

This interpretation of a religious grounding for economic development was used by some to justify Eurocentric ideas about the social bases of modernization. 'Calvinistic Protestant ethics, which promoted economic growth, was contrasted with other religions which impeded such development. Other religions often operated as a barrier to economic growth, a psychological impediment to development' (Raghuram 1999: 235). The use of Weber's work in this way fails to recognize that he stressed the importance of the context in which he developed his ideas. He did not claim to provide a universal explanation for capitalist economic development, choosing instead to consider the nature of German society in the nineteenth century (Preston 1996).

Ethnodevelopment

In 1996 Björn Hettne claimed that, 'Ethnicity has been a neglected dimension in development theory' (Hettne 1996: 15), arguing that both modernization theories and classical Marxist theories fail to consider ethnic diversity in great detail. 'Ethnicity' is a highly complex concept, but simply stated,

> 'ethnicity is seen as both a way in which individuals define their personal identity and a type of social stratification that emerges when people form groups based on their real or perceived origins. Members of ethnic groups believe that their specific ancestry and culture mark them as different from others.'
>
> (Johnston 2009: 214)

For modernization theory, many forms of cultural practice are regarded as an obstacle to development (see above) because they are considered to represent non-rational, collective, traditional ways of life, unsuited to individualistic capitalist development. For Marxists, identification by ethnicity is viewed as 'pre-modern' and a barrier to development because of the perceived irrationality of allegiance to an ethnic group, rather than a class grouping.

Regardless of ideology, the focus on development at a national scale means that any diversity within the population needs to be incorporated into a national project. Ethnic diversity can represent a challenge to this national approach. For example, Rodolfo Stavenhagen (1996: 2) states,

From the perspective of the modern nation-state, the existence of ethnically distinct 'sub-national' groups, particularly when they are politically organized, always represents a potential threat, a destabilizing force. This is especially the case when power in the state rests principally with a dominant or majority ethnic group, or when the national society constitutes an ethnically stratified system.

The term 'ethnodevelopment' has been used by Stavenhagen (1986) to describe development which takes into account the need to maintain ethnic diversity as development takes place. According to Hettne (1995, 1996), there are four main aspects of ethnodevelopment:

- Territorialism: the spatial concentration of ethnic groups, such that decisions about 'development' are made within a particular territory based on the resources of that particular area.
- Internal Self-determination: the ability for a particular ethnic group to control collectively its destiny within the context of a nation-state.
- Cultural Pluralism: the existence of and mutual respect for a number of cultures within one society.
- Ecological Sustainability: development should progress with no significant destruction of the natural environment which would threaten future livelihoods (see Chapter 6).

Plate 5.1 Tlacolula market, Mexico.
Credit: Katie Willis

These ideals are, however, much more difficult to achieve in practice for a number of reasons. For example, Stavenhagen's original concept was developed within the Mexican context, in particular the ways in which the rights of indigenous people living in rural environments were being disregarded by the urban-based *mestizo* (mixed race) majority. Within this context, the spatial concentration of groups enables a construction of ethnic identity linked to a particular territory, making the notion of ethnodevelopment through self-determination a logistical possibility, if not a politically-favoured one. The link between ethnicity and territory is not always so easily made.

The growth in 'sustainable tourism' or 'ecotourism' (see Chapter 6) has been used by some governments and indigenous groups to create income-generating opportunities over which group members have control, and do not involve the eradication of indigenous cultural practices.

In Taiwan, the Bunun indigenous group have a series of community-run facilities such as a hostel, restaurant and organic farm (Plate 5.2). There is also a village theatre for daily dance performances and workshops where visitors can find out more about traditional craft techniques and young members of the community can learn about their history (Bunun Cultural and Educational Foundation 2010).

Plate 5.2 Entrance to a Bunun Community Site in Eastern Taiwan.

Credit: Katie Willis

Heather Zeppel (1998) provides an overview of what she terms 'indigenous tourism', using examples from throughout the world, both North and South. The key to successful projects in both economic and cultural terms, is that the 'host community' of indigenous peoples is able to set limits on what is done, how it is done and when it is done (Table 5.1). For example, the Zuni Indians of New Mexico, USA limit tourist access during particular religious ceremonies, such as the summer rain dances (Mallari and Enote 1996, in Zeppel 1998), and the Aboriginal groups in the Northern Territories of Australia encourage tourists not to climb Uluru (also known as Ayers Rock) because of its sacred significance.

The role of ethnicity in framing understandings of development and contributing to development policies has grown since the 1980s, not least because of postcolonial and postmodern approaches in theoretical terms, but also the rise of campaigning groups, particularly around the rights of indigenous people. Mobilization of a rights discourse (see below) has achieved some success, but it should be noted that ethnic diversity is more than an indigenous/non-indigenous distinction with ethnic minority groups throughout the world often experiencing disadvantage and marginalization.

Development projects and policies at a number of scales have increasingly incorporated the perspectives of ethnic minority groups or indigenous populations, particularly in relation to conservation of forests or certain wildlife species (see, for example, Nursey-Bray 2009 on turtle-hunting along the Great Barrier Reef, Australia). Despite such good intentions, such attempts are often fraught with problems. These are partly because of the potential conflicts between

Table 5.1 *Dimensions of sustainable indigenous tourism*

Controls	Examples
Spatial limitation	Host community decides who can enter indigenous land and which areas, in particular sacred sites, are 'off limits'
Activity limitation	Host community establishes in what activities tourists can engage. For example, photography may not be allowed
Temporal limitation	Host community can set limits on when tourists can enter the territory. During particular religious or cultural ceremonies 'outsiders' may be excluded
Cultural limitation	Host community decides what information about cultural practices and rituals can be provided for tourists

Source: adapted from Zeppel (1998: 73)

ethnic group self-determination and the wider national project, in particular over the use of resources. Second, ethnodevelopment is based on a rigid notion of ethnic identity with individuals claiming allegiance to a certain group having access to decision-making, while others are excluded. This has proved to be an issue in a number of resource-based national schemes, such as access to gambling profits in the US for Native Americans, or rights to hold land collectively in Mexico.

Just because there is no territorial basis does not mean that a recognition of ethnic and cultural diversity is not possible. Ideas of cultural pluralism are applicable to a range of societies and are predicated on the ideas of diversity but a lack of hierarchy. The multicultural approaches adopted in a range of countries throughout the world demonstrate a desire to embrace ethnic and cultural diversity within one nation-state, but in reality again this is not always easy. One of the countries which has been very vociferous in its attempts to recognize, rather than ignore or attempt to eradicate ethnic diversity, is Singapore (Box 5.1).

Plate 5.3   Sultan Mosque, Singapore.
Credit: Katie Willis

Box 5.1

Ethnic diversity in Singapore

The historical development of Singapore as a key port and trading post in South-East Asia has led to a very diverse ethnic population. In 1994 the population of 2.9 million was made up of 77 per cent Chinese, 14 per cent Malay and 8 per cent Indian (largely Tamil), with a mixture of other groups making up the remaining 1 per cent. At the time of independence in 1965, Singapore experienced severe racial conflicts between the three main ethnic groups. The post-colonial government of Lee Kuan Yew recognized that such divisions were not conducive to economic growth and improved standards of living, and placed ethnic harmony and constructions of national, rather than ethnic identity at the heart of its policies.

Throughout the post-1965 period there has been a focus on promoting a Singaporean identity regardless of ethnic allegiance. This has been done through National Day celebrations, songs and advertisements (Kong 1995). In addition, the successful policies to promote economic growth have meant that there are fewer reasons for antagonism than in economic crisis situations. Since 1965 Singapore has moved from a GDP p.c. of US$450 to US$49,704. In addition, Singapore is now ranked 23rd out of 182 countries in the Human Development Index, and is classified as having very high human development. Attempts to promote inter-ethnic awareness and understanding include celebrating a range of religious and cultural festivals, such as Chinese New Year, Ramadan and Hari Raya, with public holidays on the main feast days. The government is also able to use its control of a large part of the island's housing stock to prevent large clusters of a particular ethnic group. It is argued that if people from different ethnic groups live in close proximity, this will help develop understanding, friendship and harmony.

These government efforts have certainly helped the development of a society where overt inter-ethnic violence is very rare. However, it is clear that not all ethnic groups have benefited equally from Singapore's economic success. In terms of household income and education levels, it is the Malays who have tended to do less well out of Singapore's twentieth-century economic growth.

Sources: Barr and Skrbis (2008); Huff (1997); Kong (1995); Perry *et al.* (1997); UNDP (2009)

Religion and development

The role of religion in development has received increasing attention since the start of the twenty-first century, particularly in relation to development practice (Bradley 2009; Lunn 2009; Tomalin 2006). As highlighted earlier in this chapter, religion has sometimes been viewed as an obstacle to progress and modernization, and as a set of beliefs and practices that would disappear over time. This has clearly not been the case. The growth of postcolonial approaches that recognize the cultural specificity of development definitions, combined with a growing realization of social diversity and the need for development policies to acknowledge this diversity, has led to religion and development assuming a more prominent role in development theory and practice.

The way in which concepts of development or progress are influenced by religion is demonstrated by Weber's work on Calvinist Protestantism (see above). Theocracies are the most extreme examples of how national development is framed by religion:

> Theocracy is the exercise of political power by the clergy of a
> particular religion usually . . . claiming to be acting primarily on behalf
> of a divinity and governing according to its principles and
> requirements.
>
> (Megoran 2009: 223)

While there are very few theocracies in the world today, with Iran being the most obvious example (Box 5.2), religion plays an important role in framing national development visions in many countries. For example, the United Arab Emirates (Abu Dhabi, Ajman, Dubai, Fujairah, Ras al-Khaimah, Sharjah and Umm al-Quwain) have experienced rapid economic growth since the 1970s, due to the discovery of large oil and gas reserves. However, not all of the emirates have these reserves so they have sought to achieve economic growth through tourism and other service industries. This has been particularly true for Dubai. The economic growth of the UAE is reflected within the landscape, as apartment complexes, shopping malls and office blocks are constructed (Plate 5.4), as in other regions seeking to 'modernize'. However, socially and culturally, legislation reflects the fact that Islam is the official religion. While there are variations between the emirates, this means that legislation regarding, for example, 'public modesty', marriage and the role of women, is framed by particular interpretations of Islamic teaching (Krause 2009).

In Bhutan, where Buddhism is widely practised, King Jigme Singye Wangchuck suggested the concept of Gross National Happiness (GNH) in 1972. Based on Buddhist principles of holistic well-being and harmony with the natural environment, this concept has been

Box 5.2

Theocracy in Iran

The 1979 Islamic Revolution in Iran involved the overthrow of a monarchy and its replacement with a theocratic form of government where religion (in this case Shia Islam) and the state came together. The revolution was a response to what was seen as an increasingly autocratic and unequal system, led by Mohammad Reza Shah Pahlavi. His attempts at economic and social 'modernization' were interpreted by many as a form of westernization and a decline in morals. This was reinforced by the support that the Shah received from the USA.

Following the revolution, the constitution was changed so that the Supreme Leader of the country was a religious leader. Ayatollah Khomeini, who had been prominent in the revolution, was appointed to this role in 1979. The Supreme Leader controls the armed forces and also nominates half the members of the Guardian Council, which oversees all legislation passed by parliament to ensure that it conforms to Islamic law. The Guardian Council also approves parliamentary candidates. The president and members of parliament are elected through universal suffrage, although reported electoral irregularities in the 2009 presidential elections suggest that elections are not always free and fair.

The introduction of Islamic rule following the 1979 revolution involved strict laws that were viewed as promoting morality. For example, Haleh Afshar argues that for many women from poor socio-economic groups, as well as highly religious women, the revolution was viewed as a possible liberation from periods of labour exploitation and what they saw as the promotion of women as sex objects. However, new legislation and the policing of public morality meant that after the revolution women's freedoms were increasingly restricted. This included legislation about dress, movement and also legal shifts which resulted in women's evidence being ignored unless it could be corroborated by a man.

In economic terms, poverty levels in Iran have reduced since the revolution, due mainly to government expenditure on health, education and infrastructure services, as well as subsidies on food, fuel and medicine. Presidents Rafsanjani and Khatami (1989–2005) embarked on pro-business policies and there was some privatization, but the economy is still dominated by the public sector. The importance of oil and gas in funding state expenditure is key as over 90 per cent of foreign exchange earnings come from this source. The election of President Ahmadinejad in 2005 has been interpreted as a reaction against the reformist tendencies of his predecessors. However, his re-election in 2009, in the disputed election, led to widespread protests.

Source: adapted from Afshar (1985); Menashri (2001); Salehi-Isfahani (2009).

Plate 5.4 Sharjah skyline.
Credit: Kelly Carmichael

adopted by the country. It encompasses the desire to improve living standards through modernization, without destroying the environment or viewing people purely as economic actors (Gross National Happiness 2010). The principal components of the GNH have been adapted in the creation of the Happy Planet Index discussed in Chapter 6.

At a sub-national level, faith-based organizations (FBOs) or faith-based development organizations (FBDOs) have increasingly been the object of research and policy attention. This links to the debates around NGOs highlighted in the previous chapter. First, such organizations are seen as working effectively with local populations for whom faith is a key element of their lives (Bradley 2009). This means that they will be able to implement programmes that are appropriate to local needs. Second, FBOs and FBDOs are also seen as being efficient providers of services and support because of their social capital (see Chapter 4). Rather than having to set up new networks, it is argued that FBOs can mobilize existing contacts. However, as Tamsin Bradley (2009) stresses, not all FBOs are the same; for example, some organizations have missionary or proselytizing goals, which may also involve restricting services to members of that faith and denigrating non-believers and followers of other religions.

Gender and development

Modernization approaches to 'development' frequently ignored gender differences in the populations under consideration. The assumption was that as economic growth took place, the benefits of such 'development' would trickle down to benefit all sectors of society. However, this did not recognize social structures that created and exacerbated inequalities and meant that the free-flowing 'trickle-down' was blocked.

'Gender' refers to the categories of 'male' and 'female', but as well as the biological characteristics which are associated with these categories, 'gender' includes the norms and expectations regarding behaviour that are associated with men and women in particular societies at particular times. 'Gender' is therefore a socially-constructed category, and as such, changes over time and space. Within the 'development process' one of the first people to focus on the ways in which women were affected differently from men through the modernization project was Ester Boserup (1989 [1970]). She argued that as societies and economies moved from a rural, subsistence base to an industrial urban core, women were increasingly excluded, leaving them on the margins of capitalist development and its perceived benefits. This was because of the association of women with the domestic and reproductive sphere of childcare and housework, while men's roles in society were constructed as involving non-domestic activities. While production was concentrated in the home and communal land, women could combine the two, but once production was moved to a different sphere of factories and workshops, women were unable to maintain their involvement in both sets of activities. While Boserup's analysis has received a great deal of criticism, not least for her generalizations (Benería and Sen 1981), her work was key in highlighting how supposedly 'neutral' development processes did in fact have very different influences depending on gender.

The increasing gender awareness led to development organizations and governments implementing policies which they hoped would involve women to a greater degree in 'development'. For example, the United Nations declared 1975–85 the UN Decade for Women. These initiatives recognized that 'development' that excluded women could not really be termed 'development', as in many cases the approaches did not really address the root issues of gender inequality and disadvantage, or involve women in making decisions about their lives.

Caroline Moser (1993) identifies five main approaches that have been adopted in relation to women and development (Table 5.2). While she stresses that some of the categories are overlapping and the chronological description does not mean that the approaches followed one another in a clear manner, the categorization is useful for examining how an awareness of gender has been incorporated in existing development theories, or has contributed to the emergence of new ones.

Moser builds on Maxine Molyneux's work (1987) regarding practical and strategic gender interests. 'Practical' interests, or in Moser's terms 'practical needs', refer to women's needs to fulfil their current socially-constructed roles. For example, if women are responsible for housework, then practical gender needs may include access to

Table 5.2 Approaches to gender and development

Approach	Date	
Welfare	1950s onwards	Targeted women in their domestic role; women viewed as passive; projects addressed women's practical gender needs, such as food aid, health and nutrition advice
Equity	1970s	Prompted by the UN Decade for Women; aimed to address strategic gender needs by eradicating obstacles to women's advancement in public sphere; strong focus on legislative changes
Anti-poverty	1970s	Women's low status interpreted as being caused by income poverty; focus of projects on income-generating opportunities for women; no consideration of patriarchal structures of oppression
Efficiency	1980s onwards	Focus on women as channels of development; during SAPs women's paid work and domestic work intensified
Empowerment	1990s onwards	Aims to lead to significant shifts in gender relations; original focus on projects devised and run by groups of women from the South; approach increasingly adopted by Northern organizations; increased focus on incorporating men into gender and development projects

Source: adapted from Moser (1993)

drinking water because this would help their current day-to-day activities. In contrast, 'strategic' needs are those which involve a change in the present state of gender relations such as changes in legislation about women's right to own land. Different approaches will focus on combinations of practical or strategic needs.

Moser's 'efficiency approach' is a good example of how a gender dimension was incorporated into existing development theories, particularly its association with neoliberal policies (Chapter 2). For example, during the implementation of SAPs in the 1980s and early 1990s, increasing numbers of women went into paid employment (Figure 5.1). While the links between SAPs and women's increased paid employment are contested, much micro-level evidence suggest that the increased cost of living and reduced state support of this period forced women to look for paid work. While for some this represents an opportunity for women to gain status and influence in the home and their communities by their access to monetary income, many studies suggest that labour force entry is associated with increased stress and health problems, as women have to combine this paid work with their continued domestic responsibilities (Dalla Costa 1995).

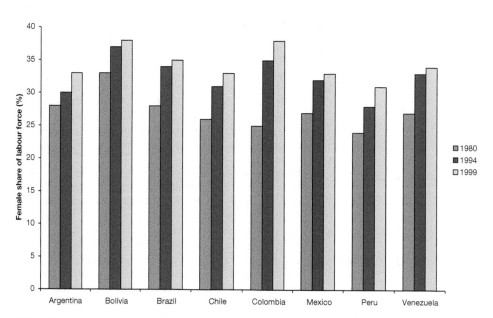

Figure 5.1 Female share of the labour force in Latin America, 1980–99.
Source: based on data from Chant with Craske (2003: 205)

Diane Elson (1995) argues that SAPs are an example of 'male bias in the development process'. This bias towards men is not necessarily intentional, but demonstrates a lack of awareness of women's lives on the part of policy-makers. SAPs are implemented at a national level and focus on macro-economic indicators (economic processes at a national level) such as government expenditure and tariff rates. However, as we saw in Chapter 2, these policies affect the lives of people at the grassroots. In gender terms, Elson argues that women are particularly affected. Because they are usually responsible for domestic activities, any cutbacks in food subsidies, increases in food prices and declining household incomes require women to make their limited household budget go further. This may mean spending more time shopping for bargains and producing food or clothing at home, rather than purchasing prepared food or factory-made clothing (see also Mullings 2009). In addition, policy-makers have often assumed that women who are not in paid work are doing nothing. In times of need, therefore, it is easy for women to 'take up the strain' and enter the workforce. This does not recognize the fact that for many women, particularly in the South, domestic chores are physically demanding and time-consuming activities. Entering paid work just

Plate 5.5 Flower packing plant, Kenya.
Credit: Katie Willis

adds to women's burdens (Chant 1994; Vickers 1991). Despite the growing use of the concept of 'empowerment' in international development policy (see below), neoliberal policies often continue to be based on the assumption of women's capacity to take on more paid and unpaid work (Chant and McIlwaine 2009: 232–5).

Economic-based understandings of gender inequality can also be seen in the ways in which socialist and communist governments have dealt with gender. Under classical Marxist theory, male privilege over women (patriarchy) is rooted in the capitalist system. According to Engels (1940 [1884]), in *The Origin of the Family, Private Property and the State*, as societies move from pre-capitalist to capitalist forms a gender division of labour becomes more apparent, with women staying in the home and dealing with domestic tasks, and men engaging in paid work outside the home. This is very similar to Boserup's arguments outlined earlier. However, Engels' arguments emphasize that capitalism and patriarchy are strongly interlinked. With women working at home providing food, shelter, clothing and childcare, the employers do not have to provide these services for their workers, so saving money. In Marxist terms, this is called the 'reproduction of the labour force'. Following this analysis, gender equality is not possible under capitalism. In order for gender equality to be achieved, therefore, capitalism needs to be replaced.

Socialist and communist governments have tended to promote gender equality, at least in their policy approaches. For example, focuses on women's education and labour force participation have been important elements of the development process in many socialist countries. Women's labour force participation has also been encouraged and facilitated by the state provision of communal services, in particular childcare provision. In some cases, communal kitchens and washing facilities have been set up to allow some women to be freed from the constraints of domestic responsibilities. Incorporating women into the paid labour force sits within socialist ideals about equality, but it can also be seen to contribute to greater efficiency within the economic system and therefore greater opportunities for economic growth (so similar to Moser's 'efficiency' approach).

Overall, women's labour force participation in socialist or communist countries is higher than in their non-socialist counterparts (Table 5.3). However, it is certainly not the case that gender inequalities are eradicated in these countries. While some women

Table 5.3 *Women's labour force participation as a percentage of men's, 1970, 1992 and 2005*

	1970	1992	2005
Non-Communist countries			
Sweden	61	92	87
Norway	38	83	87
USA	59	82	82
UK	55	75	80
Australia	42	71	80
Japan	64	69	66
Switzerland	52	62	80
Ireland	36	47	74
Ex-Communist countries			
Czechoslovakia	80	89	–
Czech Republic	–	–	77
Slovakia	–	–	76
Hungary	70	85	73
Bulgaria	79	85	78
Poland	85	82	78
Romania	83	85	80

Source: adapted from UNDP (1994: Table 34) and UNDP (2007: Table 31)

experienced a reduction in their domestic burden, the majority of women found that paid work added to an existing heavy workload in the home. State provision of communal facilities was never able to match the demand (Ashwin 2000). In addition, changing attitudes is a very difficult and long-term project. While some governments have implemented legislation to promote gender equality, such as the 1975 Family Law in Cuba, which stated that housework should be shared equally between men and women, in practice understandings of male power and dominance remain ingrained (Smith and Padulla 1996).

Moser's 'empowerment approach' does, however, suggest a form of 'development' which is more than just 'adding gender' to a pre-existing form of development theory. As outlined in Chapter 4, the shift towards a more grassroots understanding of development and approach to development implementation has provided some communities (or sections of communities) with opportunities to devise schemes that are not only appropriate to achieve improved standards of living, but also to create conditions whereby women feel able to empower themselves (Box 5.3). Gender equality and

Box 5.3

The Grameen Bank, Bangladesh

Since the early 1980s, the Grameen Bank (GB) has provided financial credit to the poorest households in rural Bangladesh. Over 15 per cent of all villages in the country are now covered by the scheme. Formal banking systems discriminate against the very poorest people in society. A lack of land or property means that these people do not have any collateral to use to gain credit. Their low and irregular earnings also make repayment difficult. To deal with these possible problems, the GB provides credit to groups, rather than individuals. This means that there is group support and pressure to repay on time. Repayment schedules are more realistic than those set by the formal banks.

While it did not set out to be a project targeted at women, the vast majority (about 98 per cent) of borrowers are women. They are particularly marginalized within the formal banking system because of low earnings and the tendency for household property to be held in the husband's name. The GB has allowed women to access funds which can be spent on developing income-generating activities such as small businesses, or may be used for household expenditure.

It is often assumed that access to income is empowering for women as it allows them a greater say in household decisions. However, this is not always the case. The GB runs various activities alongside its credit scheme. These include literacy classes and training in first aid and other welfare activities. Survey results indicate that women involved in GB activities are more likely to participate in household decision-making than non-borrowers. There are also fewer differences in provision of meals between men and women in borrower households, and women feel able to talk more openly about family planning. This is not the case for all borrower households, but it indicates that this form of grassroots activity can help processes of women's empowerment.

The Grameen Bank model has been copied throughout the world as a way of promoting women's empowerment and also a route to poverty alleviation for marginalized populations. In 2006 the Nobel Peace Prize was awarded to the Grameen Bank and its founder Muhammad Yunus. The GB has also extended its activities into other spheres, such as the provision of mobile phones that women buy or lease. They can then use them to raise income by charging neighbours (see Chapter 7 for discussion of the use of mobile phones in development projects).

Sources: adapted from Grameen Bank (2010); Hashemi *et al.* (1996); Khan Osmani (1998); Momsen (2010: 212–6) empowerment have been incorporated into international development policy, most notably the Millennium Development Goals (Number 3), but as Naila Kabeer (2005) highlights, policies must recognize the multi-dimensionality of empowerment and the diversity of women if they are to be successful.

One of the key shortcomings of many gender and development projects has been the lack of engagement with men. Far too often, men were regarded as 'the problem', but rather than dealing with 'the problem', projects continued to deal with the role of women in society and tried to help women's empowerment. This is clearly important and there are numerous examples of success throughout the world to support this. However, since the early 1990s there has been an increasing focus on incorporating men into gender and development projects and a consideration of the ways in which men's roles and positions in society are socially constructed (Chant and Gutmann 2002). These schemes have focused both on men's issues such as men's health, but also the ways in which men's behaviour has an impact on their female partners and children. It is only by allowing men the opportunity to reflect on their behaviour

Plate 5.6   Fish sellers, Essaouira, Morocco
Credit: Kelly Carmichael

that they are able to change it to the benefit of both themselves, but also the people around them.

As international organizations have become increasingly aware of how gender and development are linked, how 'development' is measured (see Chapter 1) has become more gender-sensitive. In 1995, UNDP introduced two new development measures as part of the gender and development themed *Human Development Report*: the Gender-related Development Index (GDI) and the Gender Empowerment Measure (GEM) (Box 5.4). When the GDI ranking is compared with the HDI ranking some interesting differences can emerge, reflecting different degrees of gender inequality (Table 5.4). While this recognition of the importance of gender in development experiences is significant, it should be noted that the measures used are based on 'top-down' perspectives of 'development' and also focus on the public sphere of paid employment and formal politics.

Box 5.4

Measuring gender and development

Gender-related Development Index (GDI)   Uses the same three elements of human development as the HDI, i.e. life expectancy at birth, literacy and enrolment, and income, but looks at men and women separately to calculate whether one gender has better human development levels than the other. The life expectancy figures are calculated to take into account women's higher average life expectancy due to biological factors.

Gender Empowerment Measure (GEM)   This is a measure of women's achievements in the economic and political spheres. There are three elements:

1 political participation and decision-making – measured using share of parliamentary seats held by men and women
2 economic participation and decision-making – measured by using two indicators: women and men's share of positions as legislators, senior officials and management, and shares of professional and technical positions
3 power over economic resources – measured using estimated income figures for men and women.

Source: adapted from UNDP (1995)

Table 5.4 *GDI and HDI comparisons, 2007*

	HDI		GDI		
	Value	Rank	Value	Rank[a]	HDI rank minus GDI rank[b]
Norway	0.971	1	0.961	2	−1
United States	0.956	13	0.942	19	−6
Japan	0.960	10	0.945	14	−4
Latvia	0.866	48	0.865	44	0
Brazil	0.813	75	0.810	63	0
Philippines	0.751	105	0.748	86	2
Iran	0.782	88	0.770	76	−2
Syria	0.742	107	0.715	98	−8
Kenya	0.541	147	0.538	121	3
Nigeria	0.511	158	0.499	133	0
Rwanda	0.460	167	0.459	139	1

Source: adapted from UNDP (2009: 181–4)

Note

[a] There are 27 missing entries in the GDI ranking, thus GDI ranking runs from 1–155 compared with the HDI ranking 1–182.

[b] Calculated taking out countries with no GDI ranking.

Rights-based development

Despite the ratification of the UN Declaration on Universal Human Rights in 1948, the concept of human rights has, until recently, remained outside most constructions of development theory and practice. Within rights-based approaches human rights are not just seen as a channel through which development can be facilitated, e.g. the right to a job provides income, rather, the achievement of human rights in themselves are an objective of development (Maxwell 1999).

Rights are often divided into so-called 'civil and political' rights, such as the right to vote, freedom of expression and organization, and 'economic, social and cultural' rights, including the right to food, shelter and a job (Maxwell 1999). However, this distinction has led to a range of problems with implementation. For example, it could be argued that there is a hierarchy of rights, with some rights being more important than others. Is freedom of expression really that important when people do not have enough food to live? A rights-based approach would argue that this dichotomy is not useful 'because only if people are empowered to determine their genuine needs will development occur. This . . . simultaneously promotes sustainable democracy and well-being' (Mohan and Holland 2001: 185). Rather than prioritizing particular rights, this rights-based development argues for a holistic, people-centred approach.

Different groups have increasingly drawn on the discourse of rights to make claims for the kind of development they want to occur. While talk of women's rights and indigenous rights have become commonplace, within the sphere of development, other groups have also made claims including rights based on disability and on issues around sexuality, as well as the rights of children, young people and older people (see later in this chapter). However, a problem with this form of 'identity politics' is that it can lead to rigid boundary drawing between groups and a tendency to essentialize identities, rather than recognizing them as fluid over time and space (Cornwall et al. 2008).

Globally there are about 600 million people with disabilities, approximately 80 per cent of whom live in the Global South (Handicap International 2010). Accurate figures are difficult to assess, not least because of definitional disagreements and data collection problems. For example, Philip O'Keefe (2007) highlights how the Indian census of 2001 found that 1–2 per cent of India's population had disabilities; however, using other definitions this figure goes up to 4–8 per cent. The UK Department for International Development defines disability as 'long-term impairment leading to social and economic disadvantages, denial of rights, and limited opportunities to play an equal part in the life of the community' (DFID 2000b: 2). This definition recognizes that the broader social, economic and political context within which a person lives can affect their disability, i.e. it is more than an individual impairment.

Disability is both an outcome of poverty, for example disabilities caused by malnutrition or infectious diseases, and also a cause of poverty – people with disabilities are disproportionately found among the economically poor (DFID 2000b). Limited access to education and employment, due to inappropriate facilities or discrimination, means that people with disabilities may not be able to improve their income-generating opportunities. In 2008 the UN Convention on the Rights of Persons with Disabilities (CRPD) came into force, but this does not mean that these rights are fulfilled.

The notion of 'sexual rights' has begun to be mobilized to both highlight the role of sexuality in development and also to go beyond perceived limits in identity politics mentioned above. As Andrea Cornwall, Sonia Corrêa and Susie Jolly (2008) stress, development policy has tended to ignore sexuality, apart from considering sexual and reproductive health. They explain this absence in a number of ways; sexuality is seen as private, but as Emma Tomalin (2006) argues in relation to religion, concepts of privacy have not stopped development practitioners engaging with other aspects of people's lives; Cornwall *et al.* comment that discussing sexuality can be embarrassing so it is easier to leave it off the agenda; finally, they suggest that 'sexuality and sexual pleasure [has been seen as] . . . a kind of frivolous add-on rather than something that is intimately entwined with core development concerns of poverty and marginalization' (2008: 5).

Jaya Sharma (2008) points out that 'sexual rights' refers to more than sexual minority groups, such as gay men, lesbians, bisexuals and transgender individuals, or to sex workers, but rather that the concept encompasses the sexual rights of all people. For example, the parade depicted in Plate 5.7 was a celebration of sexual diversity in Oaxaca

Plate 5.7 Sexual diversity march, Oaxaca City, Mexico.
Credit: Katie Willis

City, Mexico. While the use of items such as rainbow flags echoed 'gay pride' marches, the organizers explicitly wanted to include heterosexuals and people unhappy with fixed sexual identity labels. While sexual minority groups and sex workers may experience legalized forms of oppression and discrimination that heterosexuals may not, the concept of 'sexual rights' recognizes how sexual behaviour may be policed by the state, by religious institutions or dominant groups in society. An inclusive person-centred idea of development needs to understand 'how sexual fulfilment and autonomy can contribute to wellbeing' (Cornwall *et al.* 2008: 6).

As part of the rights-based development agenda, NGOs and other development actors have increasingly focused on working with communities to explain the concept of rights and how to go about lobbying for their rights (Gready and Ensor 2005; Green 2008). This can be seen as a form of 'empowerment' process (see Chapter 4). However, a knowledge of rights does not mean that these rights can be achieved; as Amartya Sen (1999) argued in his book *Development as Freedom*, people have to have the ability to exercise their rights.

The role of the state as guarantor of these rights is crucial. However, given the economic poverty of many countries, how can governments be expected to guarantee these rights, particularly those relating to provision of basic material needs? In addition, neoliberal policies have promoted reduced direct state involvement and growing private sector and NGO activity in the fields of economic and social development. As rights are conceived, only states are responsible for guaranteeing them, 'even if it is non-state actors (and their neoliberal policies) that caused those rights to be violated in the first place' (Manzo 2003: 437). Thus, while the focus on rights may be regarded as important for promoting opportunities for greater well-being and empowerment at the grassroots level, the implementation of such approaches is challenging.

Life stage

Another important social dimension of development is that of life stage, in particular how the needs of children, young people and the elderly are met as part of development. Just as with gender, ethnicity and culture, there has been increasing attention paid to these social dimensions in the past twenty years. Rather than spawning new theories about development, explicit consideration of the impact of age and stage in the life course on development experiences represents part of the deconstruction of 'development' as a 'one size fits all' concept and process.

Caroline Sweetman (2000: 2) sums up the focus of much development policy when she says:

> The overemphasis of mainstream development on young and middle-aged adults is understandable to some extent, since it is at this stage of life that both women and men are physically and mentally mature, become parents, and are most capable of work . . . . However . . . an enormous contribution including both paid and unpaid work is made to our societies by the young and the old. If these groups were better represented in civil service, government and development funding agencies (all institutions that replicate the age and gender biases of surrounding society), policy might reflect reality more accurately.

The increasing focus on both young and old people by development practitioners and theorists, reflects the desire to make 'development' appropriate to the needs and lives of all people. Under modernization ideas about development, not only was the route to 'development' unilinear, there was little recognition of the diversity of the population. This tendency to homogenize populations included ignoring large sectors of the population, in particular children, young people and the elderly.

Children and young people

Children as a group have not necessarily been ignored in development policies in both North and South. For example, Save the Children Fund was set up in 1919 and UNICEF, the United Nations Children's Fund, was set up in 1946. Both these organizations were set up to help children in post-war Europe, but now work throughout the world, and particularly in the Global South. What has changed is the way in which children are considered within policy-making. Whereas children were often defined as being vulnerable and in need of protection by adults, there is a growing recognition that this should not neglect the fact that children have agency and have opinions about their lives that should be listened to (Edwards 1996; Skelton 2008).

The concept of 'childhood' has been increasingly questioned. Just as the way in which societies' understandings about women's roles affect the policies implemented, so the way childhood is defined has an impact on child-related policy. Thinking of 'childhood' as being the carefree period of your life, and the time that you spend in school, fails to recognize the actual experiences of millions of people under the age of 18 throughout the world. For millions of children, day-to-day life involves paid work, housework and caring for younger siblings, sick or elderly relatives (Box 5.5). While the 'carefree' model of childhood has been criticized for being

---

Box 5.5

Young people as carers in Zimbabwe

In 1990 Zimbabwe adopted a structural adjustment programme. One of the social effects of the associated policies was increased home treatment and care for ill people who could not afford the increased costs of hospital and clinic attendance. The impact of HIV/AIDS has exacerbated this. Just as the impact of SAPs has gendered dimensions (see pp. 144–46), so there are generational impacts.

In many cases, children represent the only carer looking after an ill relative, but this is not recognized by government bodies or NGOs that work with patients with AIDS. Looking after a sick relative is emotionally draining, often physically demanding, and in many cases requires children to leave school, although there are many other reasons for school non-attendance in Sub-Saharan Africa. Young people take on the role of sole carer when there is no other adult willing or able to take on this responsibility. Even when there is assistance, children often take on some caring responsibilities. This represents a form of child work that is ignored in debate around child labour because it is unpaid and is in the private sphere of the home.

Elsbeth Robson and Nicola Ansell (2000: 180) relate the experiences of a 16-year-old boy who wrote about caring for his grandfather:

> It was already last year in September I helped my grandfather who was very ill. The first days I thought that if I help him I will be affected by such a disease. . . . As I saw that disease does not want to finished I took him and went to the hospital to be vaccinated. The doctors gave him some vaccines, but there is no change. I helped him in many ways. I went to his home and I cooked delicious meal every day. His clothes I took and washed them. Sometimes I gave him the vaccines that he had given to the hospital. After each supper I gave him some vaccines and I go to prepare him to slept.

Sources: adapted from Robson (2000, 2004); Robson and Ansell (2000)

Eurocentric, as it is based on Western ideas of what children's lives should be like, it must be stressed that many children and young people in the North have lives which are very different from the childhood idyll that is often presented. For example, Tess Ridge (2002) considers the experiences of children living in poverty in the UK. While the nature of 'poverty' in this case is very different from that in the South (see Chapter 1), issues of social exclusion and paid work as an economic necessity are apparent.

The issue of child labour is a particularly complicated one in development policies and is a good example of how policies need to be designed in particular contexts and based on the participation of groups that will be affected. In 2004, the ILO estimated that there were 190.7 million children between 5 and 14 working in the Global South, representing 15.8 per cent of the 5–14 age group worldwide. However, in Sub-Saharan Africa the figure was 26.4 per cent and in Asia and the Pacific it was 18.8 per cent. The ILO identifies particularly hazardous work that is threatening to physical and mental health, safety and moral development. In 2004 there were 74 million children aged 5–14 who were classified as working in such jobs (ILO 2006: 6).

For many people in the North, the image of young children working in sweatshop conditions is particularly distressing and a number of high-profile campaigns targeting multinational clothes manufacturers have highlighted the use of child labour, leading companies such as Gap and Nike to agree not to source clothing from factories using child labour. While such campaigns are laudable, they do not necessarily have the desired effect, i.e. children being able to go to school or stay at home, rather than earn a small amount of money working (White 1996). In addition, this form of employment represents a very small percentage of total child labour; the majority of child workers are engaged in agricultural and urban informal activities (Lloyd-Evans 2008).

In 1989 the UN Convention on the Rights of the Child (CRC) was adopted and has been ratified by all UN members except the USA and Somalia (Table 5.5). In relation to work, the Convention does not outlaw children (defined by the Convention as being aged under 18) from working, but rather recognizes children's 'right to be protected from economic exploitation and from performing any work that is likely to be hazardous or to interfere with the child's education, or be harmful to the child's health or physical, mental,

Table 5.5 *Summary of UN Convention on the Rights of the Child*

Children have the right to:

- life;
- a name;
- a nationality;
- know and be cared for by parents as far as possible;
- live without discrimination;
- education;
- health;
- benefits from social security;
- have his/her views taken into account in decisions that affect him/her;
- freedom of expression;
- freedom of thought, conscience and religion;
- freedom of association and peaceful assembly;
- protection from abuse and exploitation;
- an adequate standard of living;
- rest and leisure;
- be protected from work that is harmful to their health and development.

Source: adapted from CRC in OHCHR (2010)

spiritual, moral or social development' (CRC Article 32). The importance of listening to children and young people is a significant part of the Convention, and this principle has been increasingly incorporated into government and NGO policy. For example, the *Egypt Human Development Report 2010* has the subtitle 'Youth in Egypt: Building our Future' (UNDP and Institute of National Planning Egypt 2010). As well as including information about young people, it also focuses on the problems identified by young people themselves. However, in Egypt as elsewhere, how this information will be used and how participation can go beyond a limited form of consultation is much harder to determine (see Chapter 4).

Older people

Although the elderly, being adults, are not viewed like children as unable to make decisions because of a lack of maturity, in many cases, as Sweetman's earlier quotation highlighted, the needs of older people have often not been considered. This is largely because they are viewed as being too old to participate in work activities (both paid and unpaid). This clearly ignores the significant contributions made by people over 60. Discussions of ageing populations tend to focus on the Global North. While in percentage terms, older people make up larger proportions of the population in OECD and Eastern and Central European countries (Figure 5.2), about two-thirds of those over 60 live in the Global South (Beales 2000). Over time, these groups of older people will become more significant. For example, by 2015 it is estimated that the percentage of people aged 65 and above in East Asia and the Pacific will be 8.8 per cent compared with 7.1 per cent in 2005 (UNDP 2007: 246).

Older people often experience higher levels of poverty than younger people, due to retirement or reduced ability to undertake paid work. In addition, deteriorating health is often associated with the ageing process, but access to appropriate healthcare may not be forthcoming if health policies focus on other health issues that are regarded as more pressing (Desai and Tye 2009). The role of older people in caring for grandchildren and others is also key and can help promote feelings of usefulness and status. However, in some cases, the support, both financial and emotional, that older people may be able to provide, can result in excessive demands on their time, energy and money (Box 5.6).

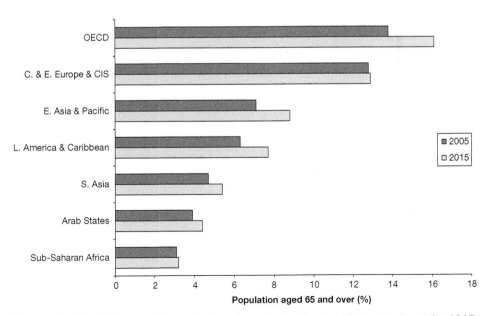

Figure 5.2 Population aged 65 and above by global region, 2005 and predicted for 2015.
Source: based on data from UNDP (2007: 246)

Box 5.6

Older people and poverty in South Africa

In 2005 3.6 per cent of the South African population were aged 65 and over, and by 2015 this figure is predicted to be 5.5 per cent, compared with a predicted Sub-Saharan African average of 3.2 per cent. These figures demonstrate that South Africa has one of the most rapidly ageing populations in Africa.

Overall levels of household poverty do not change greatly with age (see Figure). However, using the idea of trajectories of poverty, households containing older people tend to be more economically vulnerable. In times of crisis they have few assets such as land or property to help them maintain their living standards. They may also have to continue in paid work long after they would like to retire. In South Africa as a whole poverty is disproportionately concentrated within the Black African population. This pattern is also evident among older South Africans; approximately 90 per cent of the chronically poor are Black Africans, while this ethnic group only makes up 77 per cent of the population.

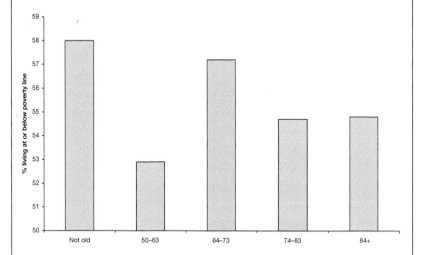

Figure 5.3 Percentage of population living at or below the poverty line by age, South Africa, 1998/9

Source: based on data from May (2003: 17)

Unlike the situation in most of Sub-Saharan Africa, old age pensions are available to many in South Africa. This money, however, is often shared with many other household members and there are examples of individuals receiving pensions acting as a magnet to others who have no income sources. Thus, older people may feel obliged to share their income with unemployed kin or to support children of other family members. Children whose parents have died of an AIDS-related illness, may also be taken in by grandparents.

Older people often feel physically vulnerable because of failing health and what they perceive as changing value systems in society. They may also be victims of abuse, not only in terms of physical mistreatment, but also through psychological abuse and financial exploitation. As populations continue to age, it is important to recognize the particular problems that older people face.

Sources: adapted from May (2003); UNDP (2007)

Given the particular needs of older people, but also their contributions to the maintenance of households and communities, and also to their countries, it is important that their opinions are considered in development policy. While the inclusion of older people in these processes has been slow, at least in terms of development rhetoric, there has been some progress. For example, United Nations declarations and conferences have increasingly incorporated ideas of older people's participation and rights (Beales 2000). In April 2002 the Second World Assembly on Ageing was held in Madrid and agreed a 16,000 word 'Plan of Action on Ageing' (Madrid International Plan of Action on Ageing, MIPAA). However, a review of the MIPAA in 2008 revealed very few concrete steps had been made to implement the Plan at either national government or multilateral organization level. In addition, the participation of older people in monitoring the MIPAA was very low (HelpAge International 2010). Thus, older people remain marginal to international development policy, while other groups are more readily incorporated.

Summary

- Modernization theories and Marxist theories view development as involving social evolution into more complex forms of social organization.
- Incorporating 'culture' into development means questioning what 'development' is and recognizing that definitions are never neutral.

● Any form of social, economic and political change will have differential impacts on individuals depending on different aspects of their identity.
● Rights-based development has been increasingly adopted as a route to incorporating social diversity into development policy and practice.

Discussion questions

1 What are the links between social and economic change in modernization theories?

2 Why are some cultural practices regarded as being obstacles to development?

3 What is ethnodevelopment and what problems are there in implementing ethnodevelopment policies?

4 Why should gender be included in a consideration of development?

5 What is a 'rights-based approach' to development and what are the potential limitations to using 'human rights' as a focus for development policy?

Further reading

Ansell, N. (2005) *Children, Youth and Development*, London: Routledge. Introductory textbook providing excellent grounding in debates around children and development.

Gender and Development (2000) 8(2): special issue on 'Gender and Lifecycles'. Excellent collection of short articles focusing on practical examples of how and why development projects should consider young people and older people in their activities.

Momsen, J. (2010) *Gender and Development*, 2nd edition, London: Routledge. A wide-ranging introductory textbook with excellent case studies.

Radcliffe, S. (ed.) (2006) *Culture and Development in a Globalizing World: Geographies, Actors and Paradigms*, London: Routledge. A wide-ranging collection of articles considering how culture and development are intertwined at different scales.

Schech, S. and J. Haggis (2000) *Culture and Development: A Critical Introduction*, Oxford: Blackwell. A very clear introduction to debates around culture and development.

Useful websites

www.bridge.ids.ac.uk/   Institute of Development Studies, University of Sussex gateway to resources on gender and development.

www.grossnationalhappiness.com   Gross National Happiness site of the Centre for Bhutan Studies.

www.handicap-international.org.uk/   Website of Handicap International. Includes discussion of UN Convention on the Rights of Persons with Disabilities.

www.helpage.org   HelpAge International. A network of NGOs throughout the world working with older people.

www.iddcconsortium.net/joomla/   International Disability and Development Consortium.

www.ids.ac.uk/go/sexualityanddevelopment   Institute of Development Studies, University of Sussex Sexuality and Development Programme. Includes links to partner organizations in the Global South.

www.ilo.org/ipec/   ILO International Programme on the Elimination of Child Labour.

www.ucl.ac.uk/lc-ccr/   Leonard Cheshire Disability and Inclusive Development Centre, University College London.

www.pathwaysofempowerment.org/   Pathways of Women's Empowerment site.

www.savethechildren.org   Save the Children Fund, USA.

www.savethechildren.org.uk   Save the Children Fund, UK.

www.survivalinternational.org   Survival International. Works for tribal peoples' rights worldwide.

www.unicef.org   UN Children's Fund.

www.worldbank.org/gender/   World Bank site that highlights how gender is dealt with in World Bank policies.

6 Environment and development theory

- Relationships between population and environment
- Modernization theory and environment
- Socialist development and the natural environment
- Intermediate technology
- Sustainable development
- Poverty and environment
- Ecotourism

Many of the theories and approaches addressed so far in the book have included implicit reference to the natural environment, but in this chapter, the ways in which 'development' and 'environment' have been considered will be at the centre of the discussion. Theories of economic growth are related to questions of resource use and distribution. Many of these resources come from the natural environment and in many cases development processes can lead to the destruction of significant parts of this natural environment.

Thomas Malthus' perspectives on population and resources

One of the earliest elaborations in the Global North of the relationship between people and natural resources was that of Thomas Malthus. In his 1798 *Essay on the Principle of Population* he wrote about the effect of rising population on the natural resource base (Malthus 1985 [1798]). While he did not talk specifically about 'development', his arguments are important for later development debates on this topic. According to Malthus, populations and food supply expand in different ways. Food supply increases arithmetically, i.e. with every generation food supply increases the same amount, by, for example, bringing new land into cultivation. This leads to a linear pattern of growth. In contrast, even if the number of children per family remains the same, the population will grow geometrically because in each generation there will be more people to have children (see Figure 6.1). As a result of these different growth rates, Malthus argued that the human population was doomed unless limits were put on population growth rates. Eventually, population would outstrip the food supply and there would be mass starvation and famine and so the population would be reduced. For Malthus, therefore, if humans did not control their reproduction, there would be disastrous consequences. In development terms, these ideas (as we shall see later) have been used to shape later development approaches in the Global South.

Malthus' work has been greatly criticized, not least because of his assumptions regarding the growth of food supply. He did not consider the ways in which new technologies may develop to increase food supply at a much greater rate. Ester Boserup (1965) highlighted how new methods and technologies can be developed to address crises such as limited food supplies in response to increasing population densities. Later technological developments, including fertilizers and new forms of seeds have been important in increasing agricultural productivity. However, just because sufficient food is produced to feed a population does not mean that everyone has access to this food. Issues of distribution are also important.

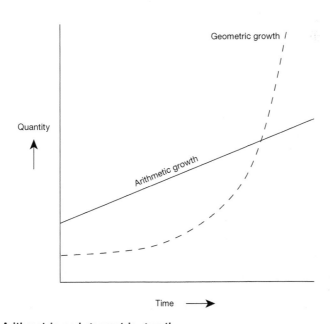

Figure 6.1 Arithmetric and geometric growth.

Environmental determinism

In Malthus' approach, the natural environment acted as an obstacle to population growth. In the environmental determinism approach, the natural environment acts not just as an obstacle, but actually shapes the nature of human society and activity. It is a form of naturalist theory as outlined in Chapter 5. Environmental determinism was popular in the late nineteenth and early twentieth centuries, and stressed the ways in which human behaviour was conditioned or determined by the physical environment.

Following this form of argument, some argued that the differing levels of prosperity, economic development or what some called 'civilization' could be explained with reference to the differences in natural environment (Huntington 1915; Semple 1911). By rooting these explanations in the natural world, some theorists argued that people from the temperate parts of the world were naturally 'better' than those from the tropical zones, and so justified the domination of Europeans over the inhabitants of other places.

As we shall see in the rest of the chapter, there is wide-ranging evidence of the ways in which human behaviour is influenced by the natural environment, but the crude theories adopted by environmental determinists are certainly out of place. They lost favour in the mid-twentieth century, not least because of the ways in which the ideas were used by certain political groups in Europe to justify racial domination. They have also been criticized because they do not consider the role of individuals, communities and governments, among others, to deal with perceived environmental constraints.

Modernization

The modernization approach outlined in Chapter 2 was built on the ideas of mobilizing technology to use resources more efficiently, not least through industrialization and the mechanization of agriculture. The basic attitude to the natural environment was one of seeing natural resources as inputs into a human-devised system. Very little, if any, attention was paid to the potential environmental damage or the long-term sustainability of such an approach.

During the Industrial Revolution in England, the environmental impacts of rapid urbanization and industrialization were clear. For example, in his descriptions of England's northern towns during the 1840s, Fredrich Engels in his book, *The Condition of the Working*

Class in England, describes the results of over-crowding, poverty and unregulated industrial processes:

> Bradford, which, but seven miles from Leeds, at the junction of several valleys, lies upon the banks of a small, coal-black, foul-smelling stream. On week-days the town is enveloped in a grey cloud of coal smoke, but on a fine Sunday, it offers a superb picture when viewed from the surrounding heights. Yet within reigns the same filth and discomfort as in Leeds. . . . In the lanes, alleys and courts lie filth and *debris* in heaps; the houses are ruinous, dirty, and miserable.
>
> (1984: 74)

Not only were these 'development' processes affecting the natural environment, they were also indirectly affecting the health of the urban populations.

Modernization and the attempts to use ever increasing areas of land for agriculture have also had severe environmental impacts. The 'Dust Bowl' of the US Mid-West in the 1930s is often used as an example of how modern technology was used to push for increased agricultural production in environmentally marginal zones. With the extension of the railways westwards in the mid-nineteenth century (see Chapter 2), large swathes of prairie land were cultivated using horse-drawn ploughs. Drought-resistant varieties of wheat were planted and farmers were able to make significant profits. However, the agricultural processes meant that during periods of drought there was often insufficient vegetation cover to protect the fine soils and high winds eroded large amounts of topsoil creating severe dust storms. With the Great Depression of the 1930s, farmers tried to increase their yields, leading to further damage (Barrow 1995; Worster 2004). Approximately 80 million hectares of grain-producing land were destroyed (Kassas 1987, in Barrow 1995). This environmental tragedy also contributed to furthering the misery of the farmers, leading many to flee the area seeking their fortunes in other parts of the USA. It did, however, trigger government action to improve policies to reduce soil erosion and help farmers use appropriate techniques.

Despite the known environmental impacts of such approaches to development, similar patterns were encouraged in the Global South; both by donor governments and agencies, and by national governments themselves. The long-term environmental problems were disregarded in favour of the goals of economic growth and development. Top-down large-scale projects such as dam building, mining, industrialization and rapid mechanization of agriculture were all promoted as suitable routes to development. The approach was very much one of 'grow now, clean up later'. Unfortunately the 'cleaning up' process is often very long and costly, if it is possible at all. Much environmental damage involves the destruction of ecosystems beyond repair (Box 6.1).

Box 6.1

Destruction of mangrove swamps in Thailand

Between 1961 and 1992, the area of mangrove forest in Thailand fell from approximately 365,000 hectares to approximately 174,000 hectares (Jitsanguan, 1993 in Bello *et al.* 1998: 189). This destruction was due to a number of factors, including factory and household pollution, logging for charcoal and shrimp farming.

Shrimp farming began in Thailand in the mid-1930s for domestic consumption, but was intensified in the 1970s as a response to a crisis in the fishing industry caused by reduced access to fishing grounds and the rising cost of fuel for boats. Growing demand for shrimps from the USA, Japan and Western Europe provided an opportunity for job creation and foreign exchange earnings. Government support for intensification led to a rapid rise in production, with the number of farms rising from 3,572 in 1980 to 15,072 in 1990 (Thailand Environmental Institute, 1997 in Bello *et al.* 1998: 189). However, these numbers are likely to be underestimates, given the potential for unlicensed farming. By 1991 Thailand was the world's leading shrimp producer.

Mangroves may be cleared for shrimp farms, or mangrove ecosystems may be severely affected by the chemicals and antibiotics used in intensive aquaculture. The clearing of mangroves not only destroys that ecosystem, but it leads to increased soil erosion resulting in increased sediment loads being deposited in the marine environment, devastating coral reefs and seagrass forests. Mangroves also have an important coastal management role, providing protection in the case of tsunamis and rising sea level.

The Thai government has sought to introduce conservation laws and controls on mangrove destruction through the use of zoning. However, environmental protection regulations are sometimes contradicted by economic development policies. There are also major problems with implementing regulations. As coastal areas have become polluted, shrimp farming has moved inland, again causing environmental problems.

Source: adapted from Bello *et al.* (1998: 187–91); Huitric *et al.* (2002); Páez-Osuna (2001).

'One-quarter of the people in developing countries – 1.3 billion in all – survive on fragile lands, areas that present significant constraints for intensive agriculture' (World Bank 2003: 59). African populations are particularly affected, with over one third of the total population living on fragile lands (Figure 6.2). Given these figures, a heavy-handed approach to agricultural modernization will lead to rapid environmental degradation and impoverishment of rural populations. Just as in the case of the US Dust Bowl, inappropriate technology has often been used in the name of agricultural progress in the South.

The so-called 'Green Revolution' of the 1950s and 1960s was a perfect example of modernization approaches to agriculture. The term was used to describe how scientific principles were applied to agricultural processes to improve yields in the South. It was clearly an attempt to escape from Malthusian limits on food supply. The main elements of the 'revolution' were high-yielding varieties (HYVs) of maize, wheat, rice and barley, as well as developments in fertilizers, herbicides and pesticides. There were some very positive results including India achieving self-sufficiency in wheat by 1980 and Indonesia moving from being a rice importer to a rice exporter (Willis 2006). However, environmentally there were problems.

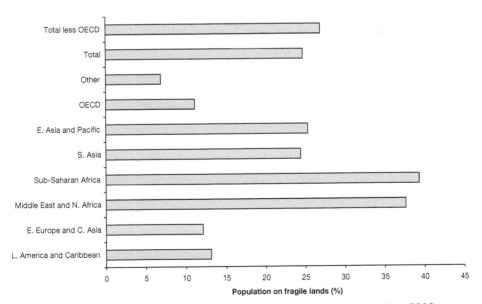

Figure 6.2 Percentage of population living on fragile lands by global region, 2000.
Source: based on data from World Bank (2003: 61)

Plate 6.1 Temporary road bridge, Ortum, Kenya.
Credit: Katie Willis

These included reductions in genetic diversity, increased demand for water because of irrigation needs and pollution from agrochemicals (Barrow 1995). In addition, the Green Revolution resulted in increased inequality as those farmers who could afford to participate reaped the benefits, while others were often forced to abandon their own land and become agricultural labourers (Shiva 1991; Yapa 1993). Current debates about genetically modified (GM) crops reflect similar positions. For some, GM crops represent a technical solution to food shortages, while for others, they are an environmental threat and will contribute to the growing dominance of the agro-chemical companies and reduced autonomy for small farmers.

The original Green Revolution techniques were adopted mainly within Latin American and Asian countries. However, the language of the Green Revolution in terms of using science and technology to improve agricultural production remains, most notably in the Alliance for a Green Revolution in Africa (AGRA). In contrast to earlier schemes, however, there is a strong commitment to environmental protection and maintenance of crop diversity, rather

Plate 6.2 Slash and burn agriculture, Sarawak.

Credit: Katie Willis

than the monocropping that was often found in the earlier period (AGRA 2010). Carol B. Thompson (2007) highlights some of the potential tensions around such aims, including the roles of transnational agribusiness, external donor expectations and national governments.

Socialist approaches to the environment

The modernist aims of many communist or socialist governments have also resulted in the implementation of development strategies which are extremely environmentally destructive. The control or taming of nature has often been a key element in the development strategies of centrally-planned economies, with rhetoric regarding the superiority of such societies being reflected in the domination of nature. According to Marx, development involved human ability to transform nature to increase standards of living. In *Capital*, Marx's perspectives on humans' dominance of nature is reflected in this description: 'He [sic] develops the potentialities slumbering within nature, and subjects the play of its forces to his own sovereign power' (1909: 283).

Some of the largest individual development projects have been implemented within centrally-planned economies. This is partly a reflection of the desire to be seen to be achieving greater infrastructure successes, but also because of the ability of governments in such economies to marshal resources to achieve these aims. The USSR provides us with a number of examples of such mega-projects, the environmental effects of which are still evident today. For example, in the 1950s the Soviet leader Nikita Khrushchev sought to increase agricultural production by bringing new lands into cultivation. The so-called 'Virgin Lands Scheme' was launched with the intention of bringing 250,000 km^2 of land into wheat cultivation in Northern Kazakhstan and Western Siberia. While cultivation was expanded and production went up, the environmental damage was enormous. Massive areas were exposed to soil erosion leaving vast swathes unusable for any purpose. Soviet attempts to increase cotton production in Central Asia also had disastrous effects, not least on the Aral Sea (Box 6.2).

Plate 6.3 Abandoned ship, Aralsk, Kazakhstan.
Credit: © Oliver Wolff/VISUM/Specialist Stock

Box 6.2

The Aral Sea basin crisis

Since the early 1960s the Aral Sea in former Soviet Central Asia has been shrinking. In 1960 it was the world's fourth largest inland water body and covered about 67,000 km^2, but by 2006 it had shrunk to 17,382 km^2 and had split in two. This is having a devastating impact on both the natural and human environments. As water levels fall and the lake bed is exposed, salt and dust are blown into rivers and irrigation systems, leading to increases in pollution and a deterioration in human health. In addition, populations earning a living from the Aral Sea are suffering as it shrinks, leaving fishing boats high and dry (Plate 6.3). For example, the fishing port of Aralsk now lies 60 km from the shore. In addition, the two smaller seas have increased levels of salinity making them unsuitable for many forms of aquatic life.

The roots of this crisis lie in the Soviet period (see Chapter 3). The Soviet ideology stressed the power of humans over nature and many large-scale environmentally-damaging schemes were adopted to further economic growth. Vast quantities of water were diverted from the Amu Darya and Syr Darya rivers to irrigate cotton. As these rivers were the main source of water for the Aral Sea, this diversion meant far lower inputs into the Aral Sea and a subsequent shrinking. Cotton acreage in Uzbekistan increased from 1.3 million hectares to 2.1 million hectares in the period 1960–80. This increase and improved yields because of irrigation led to rising Soviet cotton yields from 2.2 million tons in 1940 to 9.1 million tons in 1980. Yields in Uzbekistan are now down to 1960 levels because of land degradation and salinity problems with irrigation waters.

With the collapse of the USSR, the problem has not improved. Attempts at coordinating a strategy are limited by the fact that there are now a number of national governments involved. While Kazakhstan, Turkmenistan and Uzbekistan border the Aral Sea, the other Central Asian states of Kyrgyzstan and Tajikistan also need to be involved as the two main rivers run through their territory. The five Central Asian countries set up the International Fund for the Aral Sea (IFAS) in 1997, both to coordinate policies to protect the Aral Sea and to channel funds from external donors such as the World Bank. However, over 90 per cent of the water taken out of the Aral Sea basin continues to be used for irrigation. Despite the importance of agriculture to the economies of all five Central Asian countries, it is vital that changes are made. These could include more efficient irrigation systems as well as a move towards crops which are less irrigation dependent.

Sources: adapted from Micklin (2007); Spoor (1998); World Bank (2003)

The environmental impacts of such mega-projects are still on-going but this has not prevented the continued use of such projects in some centrally-planned countries. The most high-profile example today is that of the Three Gorges Dam across the Yangtze River in China. Despite significant evidence suggesting the potential environmental damage resulting from the project, not to mention the social problems arising from the mass relocation of an estimated 1.3 million people (IRN 2009), the project progressed. It has not been funded by the World Bank, reflecting some change in multi-lateral agency approaches to such mega-projects. However, it should also be recognized that electricity generation through hydro-electric power is much cleaner than coal-burning power stations. It is estimated that the new dam will save the annual burning of 50 million tons of coal and the release of 100 million tons of carbon dioxide (Xiong 1998, in Woodhouse 2000: 146).

Limits to growth

Modernist projects with their focus on technological solutions to perceived limitations of the natural environment were challenged by the increasing environmental movements in many parts of the world during the 1960s. A number of high-profile environmental cases in the North drew attention to the possible environmental problems which could accrue from particular forms of development. For example, in 1962 Rachel Carson's book *Silent Spring* was highly significant in drawing the attention of a Northern, particularly US, audience to the environmental side effects of certain forms of modernization. Her book dealt with the environmental impacts of the insecticide DDT, in particular the way that it was stored in organisms that ingested it, and so was passed up the food chain in larger and larger quantities, leading to the deaths of mammals and birds.

In addition, in 1972 Donella H. Meadows *et al.* published *The Limits to Growth* commissioned by the Club of Rome, a non-governmental research organization dealing with 'global problems'. The report placed the relationships between economic growth and the natural environment at the centre of the debate. However, unlike the environmental movement, which stressed the issue of environmental destruction as a problem in its own right, the Club of Rome's focus was much more on how current development methods would lead to catastrophe for the human population in terms of both rapidly declining populations (as predicted by Malthus) and huge decreases in rates of industrial growth. Meadows *et al.* stated 'we can thus say with some confidence that, under the assumption of no major change in the present system, population and industrial growth will certainly stop within the next century, at the latest' (1972: 126).

The basis for these doom-laden predications, were the results of a complex systems model which looked at five main processes: population growth, non-renewable resource use, pollution, food supply and industrialization. The relationships between these different factors and the current and predicted levels were also included in the model. By running the model with changes in the levels of the different factors, estimates could be made of when the 'limits to growth' would be reached (see Figure 6.3). The authors stressed that while the predicted levels may not be completely accurate, the overall trends were correct. These predictions about future catastrophe led these researchers and others with similar views to be categorized as 'neo-Malthusian'.

As we saw above, Malthus has been criticized for not considering the ways in which technological advances could increase the food supply. Meadows *et al.* ran their model to include a range of technological advances, such as improved mining techniques to increase access to minerals, but they still came to the same

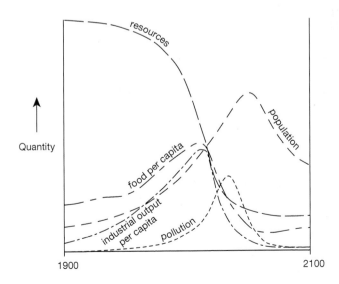

Figure 6.3 Limits to growth model.

Source: adapted from Meadows *et al.* (1972: 124)

conclusion – if current rates of consumption and economic development continued, disaster would strike before 2100. The model does not include the social dimensions of life because they are so complicated and difficult to assess. The authors are very explicit about their exclusion of these factors, but do state that decisions about income distribution, for example, could have significant impacts on when the 'limits' would be reached. In their final chapter, they state that it is crucial that decisions are made immediately about the trade-offs needed to achieve global equilibrium, i.e. not to reach the limits to growth. They state,

> As soon as society recognizes that it cannot maximize everything for everyone, it must begin to make choices. Should there be more people or more wealth, more wilderness or more automobiles, more food for the poor or more services for the rich? Establishing the societal answer to questions like these and translating those answers into policy is the essence of the political process.
>
> (1972: 181–2)

The concluding commentary from the Club of Rome executive states that it is crucial for the 'developed countries' to take a lead in this process as they are the major users of resources. Rather than stopping development in the South, resource use must be cut back in the North and attempts must be made to promote more effective and sustainable development throughout the world.

The Club of Rome report is, therefore, adopting a view of development which places economic growth at the heart of the process in terms of helping to improve poorer peoples' standards of living. However, the nature and rate of this growth must be controlled, so as to ensure that future generations have access to non-renewable energy sources, minerals and agricultural land, as well as a non-polluted environment. They do not make any specific suggestions as to the nature of policies which should be introduced, although they clearly do not suggest leaving this to market forces. Governments throughout the world are considered key in implementing policies to help reduce birth rates, conserve non-renewable resources and control pollution.

Intermediate technology

The concept of intermediate technology was developed by E.F. Schumacher and is outlined in his book *Small is Beautiful*.

The subtitle to the book is *A Study of Economics as if People Mattered* and demonstrates what he felt the focus of economics should be. Rather than concentrating on maximizing flows of money and economic growth, Schumacher argued that economic policies everywhere in the world, should be people-centred (see also Chapter 4). This was not only to allow individuals to be creative and experience the full range of what it was to be human, rather than merely a cog in a large economic machine, it was also in recognition of the environmental destruction that was occurring through the use of existing economic approaches. This environmental destruction was not only in terms of resource depletion, in particular the use of energy reserves, but it would also lead to the reduced carrying capacity of the world, i.e. the maximum number of people who could be supported using the world's natural resources.

For Schumacher the answer was not to return to a pre-industrial, 'primitive' stage; rather it was to implement policies which were appropriate to the needs of particular groups of people. In countries where there were large numbers of people without formal employment, Schumacher stated that it was bad practice to implement policies using high-tech equipment which could do the work of hundreds of people. Such an approach would result in a dual economy with the majority of the population scraping a living, while a few gained from being part of the modern capitalist economy. In contrast, Schumacher stressed the use of technology that would employ large numbers of people in productive activities, particularly in rural areas. This form of technology he termed 'intermediate technology' to highlight its position between the 'primitive' forms of tools used in the past and the very advanced high-technology equipment that had been introduced into many parts of the South in the process of development (Box 6.3).

Sustainable development

Many of the debates outlined so far in this chapter became subsumed under what has become known as 'sustainable development'. During the 1960s and 1970s the environmental impacts of various development processes were increasingly recognized by a range of groups. In 1983 the United Nations set up an independent organization, The World Commission on Environment and Development (WCED), chaired by the then prime minister of Norway, Gro Harlem Brundtland. The aim of the WCED was to

---

Box 6.3

Renewable energy: Small wind power systems in Sri Lanka

Over 80 per cent of Sri Lanka's population lives in rural areas (UNDP 2009), with most of these rural residents unable to access the mains electricity grid. Rural communities therefore tend to rely on kerosene lamps and car batteries. Charging the batteries is expensive relative to the local low cash incomes and kerosene burns are a common danger, particularly for children.

Small wind turbines have been adopted in some villages through a scheme run by the NGO Practical Action. The turbines provide cheap and safe power, which can be used to charge the car batteries and also power the lights, making the kerosene lamps redundant. Weerasinghe and his family are subsistence farmers in the village of Usgala in Sri Lanka. Following the installation of a wind turbine system, they no longer have to spend US$8 per month on charging batteries. In fact, they are able to earn a small amount of money charging their neighbours' batteries. Having electricity in the house also means Weerasinghe's children are able to do their homework after dark and therefore do better at school.

Because of Practical Action's focus on appropriate technology, villagers are trained to install and maintain the wind turbines, and local materials and labour are used to manufacture the turbine parts. In other parts of Sri Lanka, Practical Action are also using similar principles in introducing small-scale hydropower and solar power schemes.

Source: adapted from Practical Action (2010)

---

examine the problems of environment and development facing the world and to consider possible solutions. These solutions should be considered not just for current generations, but with an awareness of long-term issues.

In 1987 the WCED published its findings in a report entitled *Our Common Future* (although it is also known as *The Brundtland Report* after the WCED Chair). The Report laid out the environmental challenges facing the world, and examined how environmental destruction would limit forms of economic growth, but also how poverty and disadvantage contribute to environmental destruction. The Report stressed the importance of 'sustainable development' as a goal towards which the international community should work. According to the WCED, 'sustainable development' is: 'development that meets the needs of the present without compromising the ability of future generations to meet their own needs' (WCED 1987: 43).

Building on this environmental focus, the United Nations held an international conference in Rio in 1992 to consider ways in which sustainable development could be achieved. Since then, 'sustainable development' has become a key element in development theorizing and policy-making. However, the term's meanings are highly debated. As Jenny Elliott (2006: 10) argues, 'the attractiveness (and "the dangers") of the concept of sustainable development may lie precisely in the varied ways in which it can be interpreted and used to support a whole range of interests and causes.'

A broad distinction can be made between 'light green' or 'technocentric' approaches to the relationship between humans and nature, and 'dark green' or 'ecocentric' approaches, although the boundaries between the two are certainly not clearly defined (O'Riordan 1981; Pepper 1996). In technocratic approaches the focus is on humankind and the improvements in human standards of living and quality of life. In general, these approaches do not involve

Plate 6.4 Community tree nursery, Marich, north-west Kenya.

Credit: Katie Willis

radical changes in the current economic and political systems, rather a technical approach is adopted. This may be in the form of improved industrial or energy-generating systems which reduce pollution for example. Other technocratic solutions would include changing resource management procedures, for example by using market mechanisms to regulate human-induced environmental problems (see below).

In contrast, dark green or ecocentric approaches start with the premise that it is the Earth which is much more important than ideas about human progress and rapid economic growth. Because of this, the approaches are much more radical and call for massive shifts in the economic and political structures. In particular, there is a focus on much smaller-scale, local forms of organization similar to Schumacher's ideas of 'small is beautiful'. For economically richer countries and groups, the ecocentric approach would involve a huge reduction in consumption.

The Happy Planet Index (HPI) is an attempt to evaluate the balance between environmental destruction and human development at a global scale. It uses measures of life expectancy at birth, life satisfaction and ecological footprint. The latter is the 'the amount of land required to provide for all their resource requirements plus the amount of vegetated land required to sequester (absorb) all their CO_2 emissions and the CO_2 emissions embodied in the products they consume' (Happy Planet Index 2010). Calculating the HPI gives an indication of the ecological cost of long and happy lives. National figures provide a different perspective on global development from that provided by more conventional measures (Figure 6.4).

Climate change and development

Since the rise of sustainable development as a focus of development policy, the agenda has been increasingly dominated by the challenges that climate change poses to development, particularly in relation to poverty alleviation in the Global South. For example, the 2007/8 *Human Development Report* was subtitled 'Fighting climate change: Human solidarity in a divided world' (UNDP 2007) and the *World Development Report 2010* was on the theme of 'Development and climate change' (World Bank 2010e).

While the impact of climate change is impossible to predict exactly, due to the complexities of the global climate system, there is

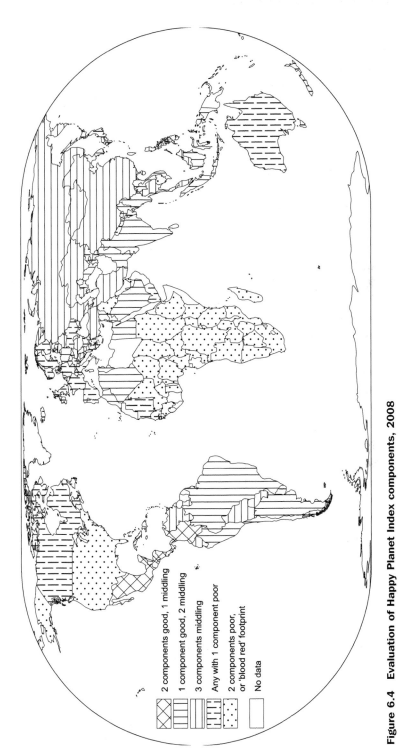

Figure 6.4 Evaluation of Happy Planet Index components, 2008

Source: based on data from Happy Planet Index (2010)

Legend:
- 2 components good, 1 middling
- 1 component good, 2 middling
- 3 components middling
- Any with 1 component poor
- 2 components poor, or 'blood red' footprint
- No data

widespread agreement that countries of the Global South will be disproportionately affected. For example, food security will be threatened due to more irregular rainfall, as well as possible increases in crop destruction due to pests. More extreme weather events, such as tropical storms and hurricanes, could lead to crop destruction, as well as loss of life and infrastructure damage (Adger *et al.* 2003). Duncan McGregor (2008) outlines the impact of droughts and hurricanes on Jamaican agriculture, suggesting possible future food security issues that may be faced by many Southern countries.

Sea level rise also represents a threat, with estimates ranging from 18 to 59 cm by the end of the twenty-first century (Huq *et al.* 2007). Thus areas currently at risk from coastal flooding would increase, as would the areas at risk during storm surges. The survival of some countries, most notably low-lying island states in the Indian and Pacific Oceans, such as the Maldives and Kiribati, is under threat even from moderate sea level rise, while large sections of the population in many countries are found in the low elevation coastal zone (LECZ), which is less than 10 m above sea level (Table 6.1).

Responses to climate change have been classified as mitigation, which is attempting to limit future climate change through measures to reduce greenhouse gas emissions, or adaptation, is adjusting to current expected climate change (Ayers and Dodman 2010).

Table 6.1 *Population in the Low Elevation Coastal Zone, 2000*

	Top ten by total population in LECZ			Top ten by share of population in LECZ	
	Number (000s)	% of total population		Number (000s)	% of total population
China	143,880	11	Bahamas	267	88
India	63,188	6	Surinam	318	76
Bangladesh	62,524	46	Netherlands	11,717	74
Vietnam	43,051	55	Vietnam	43,051	55
Indonesia	41,610	20	Guyana	415	55
Japan	30,477	24	Bangladesh	62,524	46
Egypt	25,655	38	Djibouti	289	41
USA	22,859	8	Belize	91	40
Thailand	16,478	26	Egypt	25,655	38
Philippines	13,329	18	The Gambia	494	38

Source: adapted from McGranahan *et al.* (2007: Table 3)

In both cases, the people most affected by climate change are the least able to do anything about it. While mitigation has tended to focus on international negotiations (see next section on global governance), it does also include measures to adopt clean technologies, such as solar power (see Box 6.3). Adaptation measures include early warning systems, the use of drought and pest-resistant seeds and the strengthening of coastal defences. However, as Jessica Ayers and David Dodman (2010) argue, these tend to be technocratic solutions that do not take into account the social, political and economic causes of vulnerability, which mean that certain groups are more likely to be affected than others (Wisner *et al.* 2003).

Global governance and environmental protection

Climate change is an obvious example of an environmental problem that is not restricted to within national borders. Similarly, the source of water and air pollution can be in one country, but the effects can be experienced elsewhere. Because of the global nature of many environmental problems, attempts have been made to organize responses on a global scale. The idea of 'global governance' has been used to describe political decision-making at a global level. This is not global 'government' in the sense of an elected body that represents the citizens of the world, but rather the way in which political power is exercised at this scale. Organizations such as the UN can be described as 'global governance organizations', consisting of nation-state representatives (see Chapter 7 for further discussion).

In terms of global governance and environmental protection, the 1972 UN Conference on the Human Environment in Stockholm is often held up as a key moment and in the following year the UN Environment Program (UNEP) was established (Barrow 1995). Since then there have been numerous attempts to produce global agreements on issues around pollution, whaling and biodiversity among others. The *Brundtland Report*, the 1992 Rio Conference and the 2002 Rio + 10 Conference in Johannesburg, South Africa brought these issues to even greater public attention and also highlighted the potential conflicts between North and South regarding the environmental agenda. For many Southern nations, the concept of controls on economic growth because of environmental concerns was interpreted as a way of limiting development progress in the South by denying access to methods that were used by Northern countries in their industrialization processes.

When the global distribution of greenhouse gas emissions is examined, it is clear that emissions are disproportionately concentrated in the industrialized nations of Western Europe and North America (Table 6.2). For example, while the USA makes up about 4.6 per cent of the world's population, it contributes 22 per cent of global carbon dioxide emissions.

China's role as a major carbon dioxide emitter from the Global South needs to be recognized; it also contributes nearly 20 per cent of the global figure, but with a significantly larger population, meaning that China and the USA have very different per capita emission figures; 3.9 metric tons p.c. in China and 19.7 metric tons p.c. in the USA (World Bank 2010e: 362). Attempts to control future emissions are challenging, not least because of these distribution patterns and demands from countries in the Global South to be able to develop and achieve economic growth as Global North countries did. Such

Table 6.2 *Energy consumption and carbon dioxide emissions by region*

| | Electricity consumption Kilowatt-hours p.c. | | Carbon dioxide emissions | | |
| | | | Metric tons p.c. | | Share of world total % |
	1980	2004	1980	2004	2004
Developing countries	316	1,221	1.3	2.4	42.5
Least developed countries	59	119	0.1	0.2	0.5
Arab States	489	1,841	2.8	4.5	4.7
East Asia and Pacific	253	1,599	1.4	3.5	23.1
Latin America and Caribbean	845	2,043	2.4	2.6	4.9
South Asia	132	628	0.6	1.3	6.7
Sub-Saharan Africa	463	478	1.0	1.0	2.3
Central and Eastern Europe and CIS	–	4,539	–	7.9	10.9
OECD	4,916	8,795	11.0	11.5	46.0
High-income OECD	5,932	10,360	12.6	13.2	41.9
High income	5,873	10,210	12.6	13.3	44.8
Middle income	583	2,039	2.3	4.0	42.0
Low income	106	449	0.4	0.9	7.2
World	**1,444**	**2,701**	**4.3**	**4.5**	**100.0**

Source: adapted from UNDP (2003, Table 19, pp. 300–3) and UNDP (2007, Table 22, pp. 302–5; Table 24, pp. 310–13)

tensions were very evident at the UN Climate Conference in Copenhagen in December 2009 (see Box 6.4). However, as highlighted above, the need to address the negative effects of global environmental change is particularly acute in parts of the Global South where sea level rise, land degradation and the spread of disease threatens international attempts at poverty reduction.

The limited outcomes of the Copenhagen Conference should not, however, hide the fact that international agreements on environmental protection can be successfully implemented. For example, the 1987 Montreal Protocol led to reductions in the manufacture and use of chlorofluorocarbons (CFCs), which were implicated in the development of holes in the ozone layer that protects the Earth's surface from particular forms of ultra-violet radiation (Barrow 1995).

Pricing the earth

Despite the overall acceptance that the natural environment needs to be considered as part of development policies, the section on sustainable development shows us that there are very different ways in which sustainable development is conceived. This is reflected in the range of policy suggestions. Given the problems with implementing radical 'ecocentric' approaches, it is not surprising that governments throughout the world have focused on technocratic solutions to perceived environmental problems (Barrow 2006).

Within a free market system, environmental controls may be regarded as limits to free trade, or providing too great a brake on potential economic growth. This does not mean that controls or restrictions are not applied as in the case of CFCs and the Montreal Protocol. For example, there are international agreements on the trade in hardwoods and endangered species because of the recognition that allowing free trade in these goods would result in reduced biodiversity and other potentially damaging environmental impacts. In practice, while these restrictions work to some extent, there are still significant examples of these rules being flouted. In some cases, the rule-breakers are individuals who operate without the knowledge of the law enforcement authorities, but in others, rule breaking may be ignored by the authorities, who see the need for foreign currency as being much more important than the environmental protection agenda.

Box 6.4

The United Nations Climate Change Conference, Copenhagen 2009

The UN Climate Change Conference held in Copenhagen, Denmark in December 2009, was billed as the 'last chance' for international leaders to make an agreement regarding climate change. The two-week summit involved 115 world leaders and their negotiating teams, as well as representatives of thousands of civil society organizations, some of whom were able to act as observers.

An existing international document, the Kyoto Protocol, had been agreed in 1997 by 113 countries, but a number of key countries refused to ratify it, including the USA and Australia. In both cases, national governments argued that the emissions targets that the Kyoto Protocol involved would hamper domestic economic growth and would be unpopular with the electorate. This led to widespread criticisms, particularly from Global South countries, that the Global North was failing to respond to a global crisis that they had largely created, but that would affect the Global South most seriously.

With the Kyoto experience as a backdrop, tension between North and South was likely at Copenhagen. Kyoto was based on different expectations for 'developed' and 'developing' countries. Attempts to get rid of this distinction at Copenhagen to consider emission targets more generally, was blocked by Southern nations. However, the role of the BRIC economies, particularly China, as major greenhouse gas emitters also needs to be acknowledged.

Overall, the Conference did not deliver the agreement that was anticipated. While it was agreed that there is a need to keep global temperature rises at 2°C or less, there were no commitments to reducing emissions. Many countries that are likely to be most affected by climate change, not least the low-lying island states and Sub-Saharan African countries, had wanted the figure to be 1.5°C or less, but this was not accepted. Targets for emission levels by 2020 were included (based on previously-agreed levels), but targets for 2050 were not, despite the importance of this longer-term perspective. It has been argued that China was determined that the 2050 targets would not be included, as they would be required to participate.

The Copenhagen Accord did include commitments to providing poor countries with US$30 billion per year from 2010–12 and US$100 billion per year after that until 2020. This money is to be used for adaptation activities. There was also an agreement on forest protection, which involves paying governments to protect forests.

Source: adapted from Elliott (2006); Lynas (2009); Vidal et al. (2009); J. Watts (2009)

Another approach to environmental protection within a free market model is the attempt to put a price on nature, or on environmental destruction, so creating a market for these goods. Under this approach, price reflects the cost in market terms, rather than a less tangible idea of value, for example in spiritual terms. The introduction of market price mechanisms into environmental protection has been part of what James McCarthy and Scott Prudham have termed 'neoliberalizing nature' (2004).

The pricing of ecosystem services has expanded greatly with market-focused 'solutions' to environmental problems. Rather than seeing aspects of the natural environment purely in terms of the commodities, for example wood, water, animal products and the prices that could be obtained for them, the concept of ecosystem services recognizes the wider benefits of the environment, for example as carbon sinks in the case of forests, or biodiversity in relation to animal and plant species. The price that is charged for protecting the environment should therefore include these broader services that nature provides.

Plate 6.5 Sign against charcoal production, Mulanje, Malawi.

Credit: Katie Willis

Payment for ecosystem services (PES) (also called payment for environmental services) has been introduced as a way of promoting environmental protection while also providing local communities with a source of income (Pagiola *et al.* 2005). For example, PES has been widely adopted in Mexico in relation to forest protection (Kosoy *et al.* 2008). However, for successful adoption it is vital that the process whereby community members are able to participate is relatively straightforward. Nicolas Kosoy, Esteve Corbera and Kate Brown (2008) found that farmers in the Lacandan rainforest in southern Mexico were confused by the process, or applied to participate although they were not eligible. PES schemes also need well-resourced verification processes.

Poverty and environment

The relationship between poverty and the environment is a complex one, but it is clear that there are some connections. Poor people are often forced to live in environmentally-fragile or degraded areas. In cities, these locations may include unstable hillsides, areas prone to flooding and pollution, as well as lacking basic infrastructure such as drinking water (Plate 6.6). These poor environmental conditions may

Plate 6.6 Squatter housing, Melaka, Malaysia.
Credit: Katie Willis

lead to health problems, such as respiratory diseases or water-borne infections, which in turn can affect individuals' ability to earn a living, so exacerbating their economic and social vulnerability. In addition, people living in poverty can often not afford to improve their local environment and in many cases may be forced to contribute to environmental degradation through, for example, using local forest resources for building materials and fuel (McGranahan 1993) (Box 6.5).

These inter-connections demonstrate that environmental protection measures are about more than just the natural environment; rather, attempts at sustainable development need to be placed within the much wider context of poverty alleviation, meaningful community participation in decision-making and a recognition of the importance of social and cultural contexts (Elliott 2006; WCED 1987). These complex relationships were clearly highlighted by

---

Box 6.5

Household environmental conditions in Lagos, Nigeria

Lagos is one of the largest cities in Africa, with a population of about 9.6 million. Due to the rapid growth of the city and the inability and unwillingness of successive city governments from the period of British colonialism, through independence in 1960 to the present day, the majority of residents live in poor conditions. Modern urban infrastructural services are concentrated in the commercial district and the wealthier residential areas.

In Greater Lagos only 9 per cent of the population has access to piped water. The remaining 91 per cent of the population rely on purchasing water at high cost. Over two-thirds of cases of childhood illness have been attributed to the lack of access to safe drinking water. Given its coastal location and its expansion on a series of islands, many parts of the city are vulnerable to flooding, including low-income houses that are built on stilts on the Lagos lagoon. The lack of an adequate drainage system means that during heavy rain over 50 per cent of houses are flooded.

About 99 per cent of the population do not have access to a closed sewer system, which means that human waste is disposed of in water courses and on waste ground. Of the approximately 4 million tonnes of solid waste generated every year, only about 50 per cent is collected and disposed of in landfill sites or incineration. A small amount is recycled. Because of the poor waste collection services, households often dump their waste in the street, resulting in pollution and impacts on health.

Source: adapted from Gandy (2006); Kofoworola (2007); UN-Habitat (2008)

David Drakakis-Smith in his discussions of sustainable urban development (1995, 1996, 1997). He claimed that for sustainable urban development to be achieved five areas of urban life need to be addressed: as well as the environmental aspects, demographic, social, economic and political dimensions need to be considered.

Gordon McGranahan *et al.* (1999) stress, however, that when considering the relationships between poverty and environmental destruction, it is important to recognize the scale of the environmental issue concerned. Global warming is a problem at a global scale, although of course certain locations and populations will be more directly affected than others. In contrast, poor quality water and sanitation problems are problems throughout the world, but their impacts are felt locally. In relation to debates about urban environmental problems and sustainable cities, the general patterns are:

> The urban environmental hazards causing the most ill health are those found in poor homes, neighbourhoods and workplaces, principally located in the South.
>
> The most extreme examples of city-level environmental distress are found in and around middle-income mega-cities and the industrial cities of the formerly planned economies.
>
> The largest contributors to global environmental problems are the affluent, living preponderantly in the urban areas of the North.
>
> (McGranahan *et al.* 1999: 109)

Because of this, approaches to sustainable development will differ depending on the nature of national and local economies and societies, and political priorities.

Ecotourism

Tourism represents one of the fastest growing economic sectors in the world and can provide significant income for many Southern nations (WTTC 2010). However, tourism is also associated with severe environmental destruction as unregulated infrastructure development takes place and an area's population increases much faster than the services, such as sewage systems and local water supplies, can cope with (Simon 1997). In addition, large numbers of visitors in environmentally-sensitive areas such as mountain regions can lead to soil degradation, pollution and the disruption of local ecosystems (Sharpley and Telfer 2008).

The sustainable development agenda has led to attempts to make tourism development more environmentally friendly and this has led to the phenomenon of 'ecotourism'. Sustainable tourism in its broadest sense encompasses more than environmental protection (see discussion of indigenous tourism in Chapter 5) (Hall and Lew 1998). As with 'sustainable development', 'ecotourism' can cover a multitude of activities. It may be used to refer to tourism to areas of outstanding natural beauty or interest. In other words, the 'eco' refers not to the way in which the tourism is being organized, but the purpose of the visit such as wildlife-watching. However, even within 'ecotourist' projects that are set up to reduce the environmental impact of the activities, there are wide variations. Requiring little adaptation to the 'traditional' form of mass tourism, hotels increasingly give guests the opportunity to reduce water and energy use and pollution from detergents by using towels for more than one day. As Richard Butler (1998) argues, while this does have a small environmental impact, it also helps public relations, but it does nothing to deal with larger sustainability issues.

'Ecotourism' projects which really get to grips with the concept of minimizing the environmental impacts of tourism need to look at

Plate 6.7 Tourist facilities, Petra, Jordan.
Credit: Katie Willis

limiting numbers to control pressure on resources, what form of accommodation is provided and where it is located, and the sourcing of building materials, food and other inputs. Making the tourism experience more expensive can be a way of limiting the number of visitors and can also create a fund which can be used for local environmental projects. However, such schemes do not always result in the desired outcomes in environmental terms (Box 6.6). Once

---

Box 6.6

Ecotourism in Nepal

Every year thousands of tourists travel to Nepal attracted by the potential for mountain trekking. In an attempt to protect the natural environment, while also allowing for an important source of income for economic and social development, the Annapurna Conservation Area Project (ACAP) was set up in 1986.

In environmental terms, trekking groups contribute to deforestation through their use of wood for fuel. The National Trust for Nature Conservation (NTNC) estimates that the average trekker uses twice as much fuelwood as a local resident. Wood is also used to construct lodges and teahouses for tourists. In addition, an average trekking group of 15 people generates about 15 kg of non-biodegradable and non-burnable rubbish during a ten-day trek.

The ACAP charges entrance fees to tourists trekking in the area. As of mid-2010, the entrance fee of 2,000 Nepalese rupees was worth about US$27. These fees are then used to benefit local residents and promote environmentally-sustainable activities. There are 57 Village Development Committees in the ACA and each one has a Conservation Area Management Committee (CAMC). This committee structure is aimed at promoting local participation in decision-making with the aim of making the CAMCs the main decision-making bodies in conservation matters from 2012 onwards.

In addition to entrance fees, expenditure during visits contributes to the Nepalese economy. There are clear differences between those trekkers who travel independently and those who are part of an organized group. Group participants spend more overall, but often this is spent outside the ACAP area by the agencies that are coordinating the trek. Spending on food, accommodation and equipment by independent travellers is more likely to stay in the local area.

In social terms, the ways in which local people, particularly porters, are treated by trekking groups has caused concern, with accidents and illness being much more common among the porters than the trekkers. This is a reflection of the weight they have to carry and also the conditions in which they work.

Sources: adapted from Baral and Stern (2009); NTNC (2010); Pobocik and Butalla (1998); Tourism Concern (2010)

projects grow beyond a very small scale, it is almost impossible to prevent environmental impacts. The question is how much environmental damage is allowable in the quest for tourist income?

Local production for local markets

An alternative approach to sustainable development is to focus much more on local-level activities. Rather than expending energy on transporting large amounts of goods, in particular agricultural products, around the world, for some theorists, a more locally-based self-sufficiency approach is advocated as being much more environmentally friendly and socially sustainable (Gibson-Graham 2006). This approach clearly goes against ideas of comparative advantage and the need for specialism in production and trade to allow for greater efficiency in production. The comparative advantage arguments do not, however, consider other factors such as environmental destruction.

For many producers in the Global South, the drive is to break out of the limitations of local markets to sell products on a national or international stage. Patterns in some parts of the North, however, have moved towards more local consumption. This may not be purely to reduce the environmental impacts of transportation, but they are an important part.

A good example of such an approach is the increase in farmers' markets in parts of Western Europe and North America. While local produce markets have been very important in some parts of Western Europe, in the UK for example, consumers have often been unable to buy locally-produced goods. Instead, they purchase foodstuffs from the local supermarket, which have often been imported from all over the world (Goodman and Watts 1997). Farmers' markets are set up to allow local producers to sell their produce directly to the public, rather than through large supermarket chains. Consumers may choose to purchase fruit, vegetables, meat, bread, cheese and other products from these markets for a range of reasons. They may want to support local producers, rather than the shareholders of large supermarket chains; they may feel that the quality of the food is better, particularly as they can ask questions about production processes. Finally, a local focus may be regarded as reducing harmful environmental impacts caused by what could be interpreted as unnecessary transportation of goods thousands of miles (Holloway and Kneafsey 2000).

While such reasons are clearly important and have a solid environmental grounding, this 'retreat' to more local production and consumption could have harmful effects on producers elsewhere in the world, particularly in poorer countries. If these poorer countries have adopted outward-oriented trade-focused development policies to increase economic wealth and contribute to improvements in standards of living, what would happen if their overseas markets shrank? Currently farmers' markets and similar locally-oriented forms of trade are limited and there are no signs that such approaches will have a serious impact on world trade patterns. However, this is a small-scale example of some of the complex debates around the scale of development. These will be considered further in the next chapter, which deals with how the processes that have been termed 'globalization' have affected the ways in which development has been conceived and policies adopted.

Summary

- All development theories include reference to the natural environment.
- Many development approaches have used the natural environment as a source of wealth.
- There are limits to the natural environment, but these can vary spatially and temporally.
- Sustainable development has become a key element of many development policies, but meanings can vary widely.
- There is a relationship between poverty and environmental destruction, but the link is not always clear.

Discussion questions

1 What were Malthus' arguments about the relationship between human populations and the natural environment and why have they been criticized?

2 Given the debate about the definitions of 'sustainable development', is it still worth using the term?

3 How can market-led approaches be used to protect the natural environment?

4 Why is local production regarded as a solution to many environmental problems?

5 Can global-level agreements about the environment ever work in practice?

Further reading

Barrow, C.J. (2006) *Environmental Management for Sustainable Development*, 2nd edition, London: Routledge. A clearly-written introduction to the role of different actors in managing the environment, with a particular focus on the Global South.

Elliott, J. (2006) *An Introduction to Sustainable Development*, 3rd edition, London: Routledge. An excellent introduction to sustainable development debates in the Global South.

Environment & Urbanization (2007) Special issue on 'Reducing risks to cities from disasters and climate change', 19 (1).

Peet, R. and M. Watts (eds) (2004) *Liberation Ecologies: Environment, Development and Social Movements*, 2nd edition, London: Routledge. A wide-ranging collection of chapters focusing on the relationships between capitalist development and environmental destruction and the grassroots responses to this destruction.

World Bank (2010) *World Development Report 2010*, Washington DC: World Bank. (Available at www.worldbank.org). The theme of this WDR is 'Development and climate change'. It provides an excellent insight into mainstream approaches to climate change, stressing international cooperation and technology transfer.

World Commission on Environment and Development (1987) *Our Common Future*, Oxford: Oxford University Press. The report of the Brundtland Commission that informed policy-making about sustainable development.

Useful websites

www.agra-alliance.org   Alliance for a Green Revolution in Africa website.

www.happyplanetindex.org   Happy Planet Index homepage.

www.irn.org   International Rivers Network. Works with local communities around the world to campaign for sustainable water and energy supplies, as well as flood management. Much of their work deals with campaigning against the construction of large dams.

www.practicalaction.org   Practical Action, previously known as the Intermediate Technology Development Group. Founded by E.M. Schumacher, the organization works with poor communities in the South to find appropriate technologies to meet their needs. The website includes many examples of small-scale technological approaches which have had an enormous benefit.

www.tourismconcern.org.uk   Tourism Concern. British-based organization which campaigns for ethical, fair trade tourism. There is a strong environmental focus in their work.

www.unhabitat.org   United Nations Habitat homepage. Provides information about UN programmes in relation to human settlements with a strong focus on sustainable cities.

Globalization and development: problems and solutions?

The majority of this book has dealt with the differing definitions of 'development' and the theorizations of why 'development' has or has not taken place with the nation-state as the main unit of analysis. In Chapter 4 the focus was more on sub-national levels, with an examination of what have been termed 'grassroots' approaches and in Chapter 5 different groupings within society were examined. In this chapter we look at the global scale; in particular, the development implications of processes that have been put under the heading of 'globalization'.

Globalization

While the definitions of globalization are very diverse, the majority share the basic premise that 'globalization' involves the increasing interconnectedness of different parts of the world, such that physical distance becomes less of a barrier to exchanges and movements of ideas, goods, people and money (Dicken 2007). A spatial metaphor that is often used is the idea of a 'shrinking world', while others refer to 'time–space compression' (Allen 1995). These growing linkages have been made possible because of developments in technology, transport and communications. Because of these developments, economic, political, social and cultural activities and processes which would have been limited to a smaller scale can be more easily experienced at a larger scale. This does not mean that everything now takes place at a global scale, or that all global-scale processes are experienced in the same way, but it does mean that the ways in which 'development' is examined and promoted may be very different.

There are vigorous debates about whether 'globalization' is 'new'. For so-called globalization 'sceptics', the late twentieth-century processes of global interconnectedness are a continuation of trading links and forms of cultural and political exchange which have been going on for centuries (see Chapter 1 on colonialism). However, for others, while they do not deny that there have been global level processes for a significant period, what is going on now is both quantitatively and qualitatively different from what went before (Held *et al.* 1999; Hirst *et al.* 2009).

'Globalization' has entered into the development discourse of many governments; either because it is regarded as an opportunity to promote growth and poverty alleviation (Box 7.1) or because 'globalization' is viewed as an inevitable reality within which nations must either play the game, or lose out in the search for development (Kelly 2000). However, it is important to recognize that globalization cannot be regarded as a causal factor in development (Dicken 2004: 7); rather, within the umbrella notion of globalization there are particular processes which will have certain place- and time-specific contexts. In addition, as we have seen throughout the book, it is important to recognize that the same process can have very different impacts both spatially and in relation to diverse groups of people. A wholesale embracing (or rejection) of what processes labelled 'globalization' can provide fails to recognize what Doreen Massey (1993) has termed 'power-geometry'. The potential benefits of time–space compression are not available equally to all, and for some groups, growing interactions with people and places from the other side of the world are not necessarily desirable.

New international division of labour

Despite the fact that growing global connections are highly diverse, it is the economic aspects which have been particularly prominent in the literature and policy on development and globalization (see Box 7.1). Changes in transport, communication and technology have meant that production processes can now take place in a range of locations that were previously not economically viable. This has

---

Box 7.1

DFID White Paper 'Eliminating World Poverty: Making Globalisation Work for the Poor'

In 2000, the British Government's Department for International Development (DFID) published its second White Paper focused on what the British government would do to contribute to the Millennium Development Targets (see Chapter 1). While a White Paper is not a piece of legislation, it does provide a summary of government policy on an issue and often forms the basis for later legislation.

The White Paper shows an awareness of the contested definitions of 'globalization', stating that although its general meaning is growing interconnectedness and interdependence, this can involve increasing movement of goods, services, people and information, as well as international agreements for environmental protection and human rights. DFID believes that globalization has great potential in terms of poverty alleviation, but that a positive outcome is not guaranteed:

> Managed wisely, the new wealth being created by globalisation creates the opportunity to lift millions of the world's poorest people out of their poverty. Managed badly and it could lead to their further impoverishment. Neither outcome is predetermined; it depends on the policy choices adopted by governments, international institutions, the private sector and civil society.
>
> (2000a: 15)

There is, therefore, a very technocratic or managerial approach to globalization. As long as the appropriate techniques are adopted by a range of development actors then poverty alleviation will follow.

According to the White Paper, the correct management falls within a neoliberal free trade system. Wealth will only be created if trading between nations is allowed to flow freely, and markets operate efficiently. There are a number of sections on the reform of global organizations such as the World Bank and United Nations. However, overall the thrust of the White Paper is that the prevailing approach to development through selected state intervention, a strong role for the market and increases in civil society participation is maintained.

Source: adapted from DFID (2000a)

---

contributed to what has been termed the 'new international division of labour' (NIDL). This refers to the shift from manufacturing in Northern countries to industrial production in the South where land and labour costs are cheaper. As we saw in Chapters 2 and 3, the NICs in East Asia, Latin America and the Caribbean were able to benefit economically from this process by becoming sites for the location of

Plate 7.1 Industrial park, Wuxi, China.
Credit: Katie Willis

TNC factories. Dicken (2007) argues that this is part of what he terms a 'global shift'.

In terms of global manufacturing output, while the USA, Germany and Japan have maintained their positions in the top five manufacturing exporting countries from 1963 to 2008, there have been a number of significant shifts in the top 15 manufacturing export rankings (Table 7.1). While a number of countries have moved into the top 15 since the 1960s, it is China's rise that has been particularly impressive. In percentage terms the top two countries in 1963 (USA and Germany) had 33.0 per cent of the world total; in 2008, Germany and China at the top of the rankings had an 18.0 per cent share.

This global shift is not just happening in manufacturing. With improvements in technology and increased educational levels in parts of the South service-sector activities are also being transferred. These include data-processing operations and call centres (Dossani and Kenney 2007; Taylor and Bain 2005). India has proved a very attractive location for what is sometimes called business process

Table 7.1 *Changing patterns of global merchandise exports by value 1963–2008*

	2008		2000		1963	
	Rank	*%*	*Rank*	*%*	*Rank*	*%*
Germany	1	9.1	2	8.7	2	15.6
China	2	8.9	7	3.9	NA	NA
USA	3	8.0	1	12.3	1	17.4
Japan	4	4.9	3	7.5	5	6.1
Netherlands	5	3.9	9	3.3	9	3.3
France	6	3.8	4	4.7	4	7.0
Italy	7	3.3	8	3.7	6	4.7
Belgium	8	3.0	11	2.9	7	4.3
Russian Federation	9	2.9	NA	NA	NA	NA
UK	10	2.9	5	4.5	3	11.4
Canada	11	2.8	6	4.3	12	2.6
South Korea	12	2.6	12	2.7	NA	NA
Hong Kong	13	2.3	10	3.2	15	0.9
Singapore	14	2.1	15	2.2	NA	NA
Saudi Arabia	15	2.0	NA	NA	NA	NA
Total		**62.5**		**68.8**		**76.7**

Source: adapted from Dicken (2003: 40); WTO (2009: 12). Dicken used WTO figures for his calculations.
Note: NA = Not in the top 15 in that year.

outsourcing (BPO) with many banks, such as HSBC, moving call
centres from Western Europe and the USA to cities such as
Hyderabad and Bangalore (Plate 7.2). The highly-skilled English-
speaking workforce are employed at a fraction of the cost of their
European and American counterparts. In addition, telephone charges
have fallen greatly since the late 1990s, making this movement
financially viable. While India has become a centre for this kind of
outsourcing, the expansion of technology has given rise to plans to
expand the BPO sector in some African countries. For example, the
connection of East African countries to international fibre-optic
cables has prompted the Kenyan government to plan growth in the
BPO sector from 8,000 staff to over 120,000 by 2020. Major
obstacles include the unreliability of the electricity supplies and
perceptions of high crime rates (*The Economist* 2010b).

The NIDL has not just involved the transfer of some production from
North to South, with the associated foreign direct investment (FDI)

Plate 7.2 HSBC branch, Bangalore, India
Credit: © Ilja C. Hendel/VISUM/Specialist Stock

from Northern-based TNCs. There has also been significant South–South investment from China and India, as well as Indian and Chinese investment in the North. For example, Tata Steel, a company originating in India, now has investments in every major global region, including plants in the UK and the Netherlands.

Chinese investment in African countries experienced a rapid rise after the implementation of the 'Go out' policy by the Chinese government in 2001. This seeks to encourage Chinese companies to invest overseas by providing subsidies, cheap loans and reducing bureaucracy (Kragelund 2009). African countries were viewed as important locations for FDI for two main reasons: first their mineral and energy resources, which were in increasing demand to fuel China's economic growth, and second, the potentially large domestic market of African consumers. By 2008 there were at least eight hundred Chinese state-owned enterprises operating in Africa (Davies 2008), as well as small private-sector businesses. Indian investment has also risen sharply, particularly within the energy sector, with the Indian state-owned Oil and Natural Gas Company, for example, investing in oil pipelines in Nigeria and Sudan (Naidu 2008).

Globalization and trade

As Ricardo outlined in his discussion of comparative advantage (see Chapter 2), trade can bring great economic benefits. With the growing possibilities for rapid transportation, almost instantaneous communications and improved technologies, production of many goods can increasingly take place far away from markets. This, it can be argued, allows for new possibilities of participating in global trade and thus accessing greater opportunities for wealth generation.

Under a neoliberal agenda, the focus on free trade has been a key tenet. Rather than protecting national markets and producers, neoliberal theory promotes openness. This, it is argued, allows for a more efficient use of resources, exchange of technology and greater opportunities for economic growth. From this it follows that protectionism leads to inefficiency, higher prices and limits to economic growth. Global trade needs to operate according to particular rules. While Adam Smith's idea of the guiding hand of the market is regarded as the appropriate mechanism, in reality it is clear that this free market does not exist. Because of this, in the post-war period there have been attempts at policing world trade through global organizations. First, it was the General Agreement on Tariffs and Trade (GATT) which was set up as part of the Bretton Woods conference (Chapter 2) and then, in 1995, this was replaced by the World Trade Organization (WTO).

The WTO is made up of 153 member states (as of August 2010) and its remit is to promote the freeing up of trade between members (WTO 2010). In contrast to the GATT, the WTO has the power to enforce trade sanctions. If a member is perceived to have broken rules about protectionism then other members can force the WTO to investigate. As outlined in Chapter 2, free trade may sound as if there are no 'losers', but in practice removing protectionist measures may lead to the destruction of local industries because of competition (Plate 7.3), or environmental problems (see Chapter 6). The WTO is often viewed as the epitome of capitalist activity in a globalized world and, as a result, WTO meetings have become the focus of often vehement protest. While this has often been termed part of the 'anti-globalization' or 'anti-capitalist' movement, in reality the protesters have represented a range of views

Plate 7.3   Market selling Chinese-produced imported goods, Meatu district, Tanzania.
Credit: © Joerg Boethling/Specialist Stock

and interests, as well as adopting different forms of demonstration (Glassman 2001).

The WTO has been criticized for being too influenced by the views of Northern countries at the expense of those of the South, despite the numerical dominance of Southern nations. Constant calls for decreased protectionism have been directed at Southern nations (Box 7.2), while the USA and the European Union, for example, continue with high tariff barriers against agricultural imports and provide subsidies to national producers. However, the growing economic and political strength of key nations in the Global South, has resulted in greater deadlock in world trade talks. For example, at the September 2003 WTO meeting in Cancún, Mexico, the so-called 'Group of 21' including China, Brazil and India, acted together to withstand the pressures of the USA and the EU and refused to agree to their proposals (Stiglitz and Charlton 2005). There was a similar collapse to the Doha round of negotiations in July 2008 and agreement had still not been reached by August 2010.

Box 7.2

WTO free trade policies

The WTO was set up as a rules-based organization promoting free trade. Two examples demonstrate the operation of these rules.

EU banana imports   Under the Lomé Convention of 1975 (with later renewals) agreement was made to allow African, Caribbean and Pacific states (ACP countries) preferential access to European markets for some products. This was particularly important for agricultural products which formed a substantial part of ACP foreign exchange income. Bananas were a key product in these agreements, forming up to 60 per cent of exports from some countries, mostly in the Caribbean. Banana imports from non-ACP countries were subject to quotas or tariffs.

In 1996 the US government made a complaint to the WTO about this preferential treatment, arguing that it violated the free trade rules. The USA does not export bananas, but US companies, such as Chiquita have large-scale interests in banana production in Latin America. The EU challenged the complaint, but the WTO ruled in favour of the USA.

US steel   In 2002, the US government implemented increased tariffs on steel imports into the USA, arguing that this did not violate WTO rules because these rules allowed for emergency measures to be introduced at times of crisis. The USA argued that the 11 September attacks and the associated economic crisis justified these measures. China, Brazil, the EU, Japan, Korea, New Zealand, Norway and Switzerland all protested to the WTO that the steel tariffs were against WTO rules. In November 2003, the WTO agreed that the steel tariffs were indeed inconsistent with WTO Safeguards Agreement and called for the USA to change the policy.

Sources: adapted from Hines (2000); Thorpe and Bennett (2002); WTO (2010)

'Fair trade'

In relation to trade, the example of the Group of 21 at the Cancún summit is one example of how poorer countries and NGOs are seeking to use free trade arguments as a route to economic growth and improved standards of living. Another route is through what has been termed 'fair trade'. This is not the same as 'ethical trade', but is part of the same movement concerned with examining how goods are produced and traded and what the impacts of these processes are on producers, the environment, etc. (see Box 7.3).

Box 7.3

Definitions of trade

Ethical trade   There is no single meaning, but it generally refers to trade within which attention is paid to environmental issues, human rights concerns, animal welfare and other social issues. Michael Blowfield (1999: 754) states 'ethical trade is best thought of in terms of scope – as a term that brings together a variety of approaches affecting trade in goods and services produced under conditions that are socially and/or environmentally as well as financially responsible.'

Free trade   Free trade occurs where there are no obstacles to the free movement of goods and services. These obstacles could include policies such as tariffs, quotas and preferential treatment for domestic over foreign companies.

Fair trade   A general term used to describe trade that provides disadvantaged producers with access to markets and a fair price for their goods. It is often used in relation to perceived unfair international trading rules, which exclude Global South producers from Northern markets, or allow Northern producers to receive subsidies when Southern governments are penalized for providing such support (Stiglitz and Charlton 2005).

Fairtrade   A certification system that ensures producers are paid a 'fair price' for their goods and a premium is also paid for community projects. Producers within the Fairtrade system are required to meet a number of criteria relating to environmental sustainability and governance.

In 1989 the Netherlands was the first country to introduce a Fairtrade labelling scheme. While free market advocates seek to promote the easy movement of goods and services around the world with prices determined by the intersection of supply and demand, within the fair trade movement, the price charged to the consumer reflects a 'fair' price based on what it cost the producer. Consumers are willing to pay the higher price because this means the producer will get a reasonable return on their effort (Fairtrade Foundation 2010).

The main areas in which Fairtrade operates are in foodstuffs, particularly bananas, coffee and tea. In the UK, purchases of products with a 'Fairtrade' label have increased greatly since 2000 with purchases of over £799 million in 2009 (Figure 7.1). Globally, there are Fairtrade labelling initiatives in 23 countries, purchasing products from over 1 million producers in over 58 countries (FLO

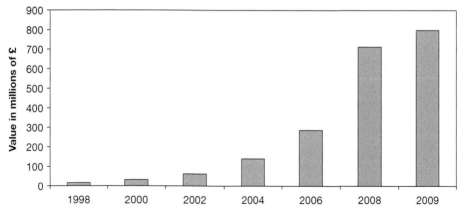

Figure 7.1   Growth in spending on Fairtrade products in the UK, 1998–2009.

Source: based on data from Fairtrade Foundation (2010)

2010). In 2007 £1.6 billion of Fairtrade-certified products were sold worldwide, an increase of 47 per cent from the previous year (Fairtrade Foundation 2010).

While Fairtrade products still represent a small but growing part of the global market, where the fair trade process has been introduced, there have been very positive outcomes (see Box 7.4). The Fairtrade example demonstrates how global linkages between and within North and South can be used to promote processes of development, not only in terms of economic growth, but also in improved quality of life.

For neoliberals however, Fairtrade policies represent an unwelcome intervention into the workings of the free market. For example, in March 2004 the Adam Smith Institute published a report criticizing the Fairtrade coffee companies (Lindsey 2004). The report argued that paying extra money for a cup of coffee just encouraged producers to keep producing coffee, when they should be diversifying into other activities. In this latter point, they are in agreement with organizations campaigning for fairer trade systems that are not biased against Southern producers, such as the Trade Justice Movement.

Regional cooperation

Inter-relationships between different countries, peoples and organizations may have increased, but this does not necessarily mean that everything now operates at a global scale. In the economic and

Box 7.4

Kuapa Kokoo Cooperative, Ghana

Since the colonial period, cocoa production has constituted an important income-generating activity in rural Ghana with thousands of small-scale farmers growing the crop to sell to 'middlemen' for sale on the global market. In the 1930s, the then colonial government took over the trading of the cocoa crops and this continued following Ghanaian independence from Britain in 1957. By doing this, the government was supposed to be able to cushion the producers from price fluctuations on the world markets. However, in the late 1970s, world cocoa prices fell by two-thirds and during the 1980s as Ghana adopted SAPs, this support was no longer possible.

In 1993, a small group of farmers set up the Kuapa Kokoo Cooperative. Rather than rely on middlemen who were known for their exploitative practices, such as weighing products incorrectly so as to underpay the producers, the cooperative decided to sell their produce directly. Some of the crop was sold to a Fairtrade company. This proved to be a successful arrangement and in 1997 the farmers decided to produce their own chocolate bar. The Day Chocolate Company was formed, with assistance from UK-based organizations. The board of the company includes farmers' representatives and all farmers have shares in the cooperative so that they receive a share of the profits. The company (called Divine Chocolate Ltd since 2007) now produces a range of chocolate bars, as well as selling Fairtrade cocoa to other companies.

The benefits of being involved in this cooperative have led to expansion and there are now around 45,000 farmer members. Not only does the cooperative help income generation, there are also important programmes of member participation and a recognition of the role of women in cocoa production. Nearly 30 per cent of the members are women. The Fairtrade system means that the cooperative receives a 'social premium', which is spent on community projects such as schools, mobile clinics, drinking water and sanitation facilities.

Sources: adapted from Divine Chocolate (2010); Kuapa Kokoo (2010); Purvis (2003); Tickle (2004)

political sphere, regional groupings have become increasingly important. Most notable in influence terms are the European Union (EU), the North American Free Trade Area (NAFTA) and the Asia Pacific Economic Cooperation (APEC) Forum. These are not, however, the only regional groupings.

According to Björn Hettne (1995), national-based development strategies in a globalizing world are increasingly difficult to implement. Frans Schuurman, in his discussion of an 'impasse' in development theory (Chapter 1) partly attributes this impasse to growing global economic interconnectedness. Within this context 'individual nation-states are assigned an increasingly smaller function. Development theories, however, still use the nation-state as a meaningful context for political praxis' (Schuurman 1993: 10). The importance of TNCs, the role of international financial institutions and greater links across national borders due to trade and migration mean that looking at development purely within the boundaries of the nation-state is untenable. The influence of external factors in a country's development status was very much part of the structuralist and dependency arguments outlined in Chapter 3. According to theorists such as Prebisch and Frank, the solution to problems of global power inequalities was to protect domestic economies either through protectionism or more extreme withdrawals from the global economic system. As we discussed in Chapter 3, these attempts have often met limited success. According to Hettne, rather than trying to operate as individual nations in a potentially hostile economic environment, Southern countries should operate in larger regional groupings.

The EU, NAFTA and APEC are some of the most obvious examples of these groupings and enable their members to benefit from being part of a supra-national organization. However, the benefits vary from organization to organization. The arguments behind regional groupings of this type are that they provide greater bargaining power compared with acting as individual nations.

Regional groupings are not a new entity within development debates. During the import-substitution industrialization (ISI) period of the 1960s and 1970s (see Chapter 3), a number of countries in Africa, Asia, Latin America and the Caribbean were involved in attempts to create regional trading agreements or common markets. The economic rationale behind such activities was to try and expand the markets for domestically-produced goods. However, just as ISI reached certain limits, so attempts at regional cooperation often faced difficulties, particularly if there were significant inequalities between members. For example, the Central American Common Market (CACM) was set up in 1960 by Costa Rica, Guatemala and Nicaragua, with El Salvador and Honduras joining soon after. Intra-regional trade increased from 6.5 per cent of the total trading of the five member nations in 1960 to 26.1 per cent in 1971, probably boosting domestic production. But political unrest, limits to ISI, followed by the debt crisis and SAPs, led to a stagnation in CACM activities (Bulmer-Thomas 1988, 1998; Grugel 1995).

The development of regional free trade groupings and institutions for cross-border economic cooperation have continued to the present day. The global situation may be different, but there is a perceived need for organizing at a level above that of the nation-state. There are a plethora of regional cooperation organizations (Table 7.2) with varying purposes. The Southern African Customs Union (Box 7.5) is an important example of how poorer countries may work together to seek advantages. There are, however, limitations regarding what such groupings may do. In trade terms, if countries are all producing the same types of goods, then the advantages of trading between each other may be limited. Within NAFTA there is a division of labour between the USA and Canada on one side, and Mexico on the other. The comparative advantages of the different countries mean that there are significant flows of goods across the borders.

As well as free trading zones or similar trade liberalization measures within the grouping, such regional organizations may play a role in maintaining political stability and security in the region. For example, the African Union is involved in peace-keeping operations in Somalia through the AMISOM (African Union Mission in Somalia), as well as being part of a joint peace-keeping force with the United Nations in Darfur, Sudan.

Transnationalism

A concept often associated with 'globalization' is that of 'transnationalism'. Transnational processes refer to sustained activities backwards and forwards across national borders (Glick Schiller et al. 1992). For example, transnational corporations consist of complex networks of research, production and marketing processes which take place in more than one country at any one time. For these corporations the transnationality of their activities is a way of maximizing profits. Production processes are often located where there is a suitably skilled and cheap workforce, as well as favourable government assistance. As argued above, individual governments are often limited in what they can do to control the movements and activities of TNCs.

Table 7.2 *Examples of regional cooperation organizations*

Name	Members	Date	Activities
African Union	53 African nations	1999 (previously Organization for African Unity OAU)	Coordination of cooperation for development
APEC (Asia-Pacific Economic Cooperation)	21 members, including Australia, China, Indonesia, Japan, Mexico, Russia, Singapore and USA	1989	Trade liberalization
ASEAN (Association of Southeast Asian Nations)	Indonesia, Malaysia, Philippines, Singapore, Thailand (1967 members); Brunei, Cambodia, Laos, Myanmar, Vietnam joined later	1967	Political cooperation on security; trade and investment liberalization
CARICOM (Caribbean Community)	15 members, including Antigua & Barbuda, Bahamas, Haiti, Jamaica, Montserrat, St Kitts & Nevis, St Vincent & the Grenadines, Suriname, Trinidad & Tobago	1965 Caribbean Free Trade Association (CARIFTA) 1973 CARICOM	Common market
EU (European Union)	27 members	1957 (European Common Market) 1992 (European Union)	Free trade area; freedom of movement; some centralized political decision-making
MERCOSUR (Southern Cone Common Market)	Argentina, Brazil, Paraguay, Uruguay	1991	Common market
NAFTA (North American Free Trade Agreement)	Canada, Mexico, USA	1994	Free-trade area. Some limitations on movements of goods and services
SACU (Southern African Customs Union)	Botswana, Lesotho, Namibia, South Africa, Swaziland	1910, although later renegotiations	Common external tariff; finance redistribution

Sources: adapted from African Union (2010); APEC (2010); ASEAN (2010); CARICOM (2010); Dicken (2007: 192); Europa (2010); MERCOSUR (2010)

Box 7.5

Southern African Customs Union (SACU)

The Southern African Customs Union (SACU) is made up of five countries: Botswana, Lesotho, Namibia, South Africa and Swaziland. It is the oldest customs union in the world, dating back to 1910, but the terms of the agreement have been re-negotiated on three occasions since then to reflect changing economic and political circumstances.

The economic power of the five members is clearly very unbalanced (see Figure). While Botswana has the highest GNI per capita figures, it is South Africa which has the most power. Within SACU there is free movement of goods, capital and services. There are, however, controls on labour mobility. The countries share a common external tariff (CET), so that it costs the same to import and export goods to and from all five countries from non-SACU members. This tariff system encourages intra-regional trade. Under the CET system, the tariff revenues are pooled and then shared among the member states. This has provided very important state revenues for Botswana, Lesotho, Namibia and Swaziland who receive a disproportionate share of the pool. According to IMF figures cited by Latham (2010), about two-thirds of Swaziland's and Lesotho's official revenues came from SACU sources.

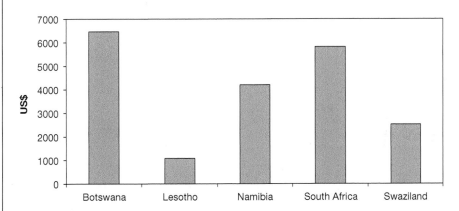

Figure 7.2   GNI per capita figures for SACU members, 2008
Source: based on data from World Bank (2010)

The future of SACU has come under increasing pressure due to the differences in economic power within the Union, as well as tensions caused by trade negotiations with external bodies. For example, the EU has sought to negotiate Economic Partnership Agreements (EPAs) with individual nations in SACU, which could undermine the Union. In July 2010, SACU leaders agreed to change the distribution of trade revenues following pressure from South Africa. Given the reliance on these revenues, particularly in Lesotho and Swaziland, this it likely to have very negative effects on the economies and populations of these countries, as well as on the future of SACU.

Source: adapted from Gibb (2004); Latham (2010); SACU (2010)

Studies of transnationalism have focused particularly on migration (Vertovec 2009). While international migration has been a feature of human activities since the world was divided into nation-states (Chapter 1), levels of migration have increased since the Second World War, partly because mobility is facilitated through advances in transport and communication technology. There are an estimated 214 million international migrants, making up 3.1 per cent of the world's population (UNDP 2009: 21). Technological changes also make it easier for migrants to continue links backwards and forwards between the host society and 'home' (Vertovec 1999). In addition, there is increasing demand for labour for work in agriculture, export-processing manufacturing and domestic service, among other sectors. When domestic labour is insufficient to meet this demand, either because of lack of numbers, or the work is regarded as too poorly paid, migrant labour will often be used (Samers 2010).

Governments of countries with large amounts of out-migration have increasingly realized the importance of using the possibilities of transnationalism to enable migrants to contribute to national

Plate 7.4 Latino murals, Santa Cruz, California.
Credit: Katie Willis

development. Remittances (the money sent to family members back home) are obviously important to the individual families and communities concerned and can result in significant improvements in standards of living (de Haas 2006). However, these flows of remittances can also represent large flows of capital into the national economy (Box 7.6). Governments may therefore introduce a range of policies to help ease the outflow of labour, even if this represents a loss of skilled workers to the domestic economy. Such policies may include government overseas recruitment agencies, the possibilities of dual citizenship and tax concessions for returning migrants.

'Hometown' associations of migrants or descendants of migrants may continue to send financial and other support back to their 'home

---

Box 7.6

The Philippines government, emigration and remittances

The Philippines is one of the most important labour-exporting countries in the world, with about 8 million Filipinos living overseas, representing about 10 per cent of the entire population (Castles and Miller 2009). While the majority of women migrate to work as domestic servants or care assistants, most men migrate to work in construction or related activities. The 'brain drain' that this outflow represents can be interpreted as a problem for development in the Philippines, but the estimated US$16.3 billion a year that is sent back to the Philippines in the form of remittances is clearly crucial to the country's economic position. In 2007, remittances were equivalent to 11.6 per cent of GDP (UNDP 2009: 160).

Government involvement in promoting out-migration has a long history in the Philippines. The Overseas Employment Development Board (OEDB) was set up in 1974 to facilitate migration and to act as an employment agency for Filipinos wanting to work overseas. In 1982 it was merged with a range of other state bodies to form the Philippine Overseas Employment Administration (POEA). Before leaving the country Filipinos must attend workshops on destinations, available support and financial matters.

Filipino workers overseas are promoted as 'heroes' by the Philippines government and a number of legal changes have meant that it is easier for Filipinos overseas to move backwards and forwards. For example, in 2003, the law was changed to allow dual citizenship and Filipinos overseas became entitled to vote in national elections. The government is keen to ensure that these migrants continue to see the Philippines as 'home', so that they will keep sending money and may eventually move back permanently and invest in domestic businesses.

Source: adapted from Castles and Miller (2009); Parreñas (2001); Ruiz (2008)

Plate 7.5 Advert for female beauty salon staff, Sharjah.
Credit: Kelly Carmichael

villages'. As with remittances, these flows can make significant contributions to infrastructural improvements such as community halls, basketball courts and health centres (Goldring 1999). The increasing significance of such flows and other transnational migrant processes need to be incorporated into our understandings of approaches to development (Davies 2010; IDPR 2010; UNDP 2009).

However, while economically remittances and other flows may be important for improvements in standards of living, it must also be recognized that there can be severe social implications of emigration. These include the breaking up of households, with parents leaving children 'at home' while they move away to work (Parreñas 2005). In addition, the working conditions which migrants face may be dangerous, unhealthy or expose them to extreme forms of exploitation (Wills *et al.* 2009). In many countries, domestic servants can only enter the country to work with a named family. This system is, therefore, like a bonded labour system whereby employees are at the mercy of their employers, unable to leave their employment because this would mean deportation (Anderson 2001).

Technology and communications

A key element of development approaches which view modernity as a goal, is technology. In particular, the way in which new technology can be adopted to help humans overcome or deal with natural obstacles, such as limited rainfall or earthquake hazard. The key reason for adopting technology is to improve efficiency, so producing more for the same effort. However, we have also seen how the drive for increased productivity using more and more advanced technologies can lead to environmental problems (Chapter 6). The introduction of new technologies can also exacerbate existing social divisions (as with the 'Green Revolution') as only some people are able to use these new technologies. This may be because of economic inequalities, but it may also reflect power distributions and norms such that certain groups, for example women, are not allowed to use particular forms of technology. The concept of appropriate or intermediate technology has been developed to address some of these issues (Schumacher 1974).

Technological advancements, particularly in the sphere of communications, have been crucial in the creation of globalization processes. The Internet, in particular, creates new possibilities for instantaneous communication and the exchange of large amounts of information between millions of people. The possibilities which such technologies have for development, particularly in the fields of education and health, have been highlighted as worthy of attention. The Internet, mobile phones, radios, computers and television can, it is argued, help overcome some of the spatial inequality problems that hinder improvements in standards of living (Unwin 2009).

For example, mobile phones can be used to find out market prices for agricultural goods, vital for farmers living in remote areas. Use of mobile phones as a channel for transferring money is also growing. The M-PESA system, set up in Kenya in 2007, has been a notable success (Plate 7.6). The system allows users to deposit, send and withdraw money using their mobile phones and by August 2009 about 38 per cent of the adult population had used the system. While the ability of internal migrants to send money to relatives is an important use of the M-PESA system, it can also be used for paying utility bills, so saving time and energy going to the payment office. It also prevents traders and taxi drivers having to carry large amounts of cash, which could make them more likely to be victims of robbery (Jack and Suri 2009).

However, it is important to recognize the significance of the 'digital divide' which exists (Table 7.3). Access to the Internet and telephone technology is not equal, with large regional differences, as well as

Plate 7.6   Sign advertising M-PESA, Eldoret, Kenya.

Credit: Katie Willis

Table 7.3   *Access to communication technology by region, 1990 and 2005*

	Telephone mainlines (per 1,000 people)		Mobile phone subscribers (per 1,000 people)		Internet users (per 1,000 people)	
	1990	*2005*	*1990*	*2005*	*1990*	*2005*
Developing countries	21	132	(.)	229	(.)	86
Least developed countries	3	9	0	48	0	12
Arab States	34	106	(.)	284	0	88
East Asia and Pacific	18	223	(.)	301	(.)	106
Latin America and Caribbean	61	. . .	(.)	439	0	156
South Asia	7	51	(.)	81	0	52
Sub-Saharan Africa	10	17	(.)	130	0	26
Central and Eastern Europe and the CIS	125	277	(.)	629	0	185
OECD	390	441	10	785	3	445
High-Income OECD	462	. . .	12	828	3	534
World	**98**	**180**	**2**	**341**	**1**	**136**

Source: adapted from UNDP (2007, Table 13, pp. 271–6)

Note: (.) Less than 0.1

spatial differences within countries and social variations, particularly along gender and age lines. The focus on such technologies to achieve development aims must, therefore, be viewed cautiously and continued attention paid to other forms of technology that do not exclude such large numbers of people. It is also crucial that organizations designing ICT-related development projects focus on the relevance of content and design for the proposed beneficiaries, rather than concentrating purely on the technology itself (Kleine and Unwin 2009).

Cultural globalization and cultural homogenization?

In Chapter 5 we saw how certain forms of 'development' have been associated with processes of eradicating particular cultural practices. These claims of cultural homogenization have been exacerbated by the increasing inter-connectedness of the world. This is especially the case in relation to consumer culture. There has been much talk of the 'McDonaldization' or 'Coca-Colaization' of the world as large corporations spread both their production centres and also their sales outlets to more and more remote parts of the globe (Klein 2000; Tomlinson 1999). For some, this spread of 'Western' consumption practices is interpreted as a form of neo-colonialism (Plate 7.7). 'Non-indigenous' music, food and clothing are promoted as being 'better' and thus those people who can afford such consumer goods are regarded as more 'developed' or 'advanced'. This sounds very similar to Nanda Shrestha's account of growing up in Nepal and being exposed to the idea that his village and way of life were 'backward' (Chapter 1).

This view of global cultural processes has been criticized, however, for failing to recognize the agency of people, communities and governments in dealing with these flows (Sharp 2009). For example, rather than obliterating existing cultural practices, there may be a process of mixing, also known as hybridization or creolization. For example, David Howes (1996) describes how Coca-Cola is used in different places stating 'No imported object, Coca-Cola included, is completely immune from creolization' (1996: 6). While its main use is as a drink, it can be mixed with a range of alcoholic beverages to produce particular local specialities. In addition, in Russia, Coca-Cola is viewed as a liquid that can remove wrinkles.

Plate 7.7 'Global' influences, Beijing.
Credit: Katie Willis

The migration of millions of peoples around the world also creates new opportunities for cultural hybridization. For example, in cities throughout the world, legacies of migration are evident in the range of restaurants available and throughout the North, the popularity of (the rather dubiously-named) 'World Music' testifies to the fact that cultural exchange is not purely a North to South process. It is clear that cultural practices and norms that were previously found only in certain parts of the world are increasingly diffused, but this does not mean that everywhere is turning into a particular vision of the United States of America.

Political mobilization

The increasingly complex networks of communication that have helped the processes of economic globalization, have also been used for purposes of political organizing. Earlier in the chapter we discussed concepts of global governance and the perceived need for global-level institutions which would regulate issues at the supra-national level. Organizations such as the World Bank, the IMF and the WTO have taken on increasing importance as the processes of globalization have developed, but activities across national borders can also be significant for smaller-scale institutions. As Robin Cohen and Shirin Rai (2000: 8) state 'a global age needs global responses'. They argue that as decisions that affect people's lives are increasingly being made at the supra-national level, it is important for responses to take place at this level as well.

Fund-raising and awareness-raising have taken on a different character within the globalizing world. For example, the Jubilee campaign to cancel 'Third World' debt is a global movement, using websites, email and text alerts (Jubilee USA 2010). The use of online petitions and blogs has also transformed the way information can be distributed and how individuals contact their elected representatives, while social networking sites, such as Facebook, have helped to create online communities of activists and supporters.

At a smaller scale, the activities of local groups can be promoted and publicized through the use of global-scale technology. Whereas in the past, local-level activities would only be known about at the local scale unless international organizations publicized them through books or leaflets, organizations can now use the technology available to 'talk to the world'. There are clearly obstacles to involvement, such as access to the technology and the reliability of telephone, electricity and satellite communication in remote areas (Table 7.3). However, there are numerous examples of organizations based in the South using such approaches to communicate with similar groups elsewhere, key players in global development decisions or the wider global public (Havemann 2000). For example, Sarah Radcliffe (2001) discusses the range of actors within the Project for the Development of Indigenous and Black People in Ecuador (with the acronym PRODEPINE in Spanish).

One of the most successful Southern-based organizations that has used new technologies to promote its message around the world is the EZLN (Zapatista National Liberation Army), also known as the Zapatistas. They came to public prominence on 1 January 1994, the day that the North American Free Trade Agreement came into force. While the demands of the EZLN for indigenous rights in Chiapas were directed at the Mexican government, the EZLN also promoted their message to the rest of the world and used their experience as a way of highlighting the plight of indigenous and marginalized rural populations throughout the world. Manuel Castells referred to the

Zapatistas as 'the first informational guerrilla movement' (2004: 72) because of their use of communication technologies. Anti-government protesters in repressive states have also used modern technologies to publicize their plight, particularly at times of public unrest, as in Burma/Myanmar in 2007 (see Box 7.7) and around the Iranian elections of 2009 (Box 5.2).

Box 7.7

The Internet and pro-democracy movements in Burma

Since 1962, Burma (also known as Myanmar) has been ruled by a military government. While there have been elections, the results have not been respected by the government and they have remained in power. For example, the National League for Democracy (NLD), founded by Aung San Suu Kyi, won elections in 1990 but there was no handover of power. The government adopts highly repressive strategies to deal with opposition and has an appalling human rights record. Aung San Suu Kyi herself was under house arrest for many years, before her release in November 2010.

While there have been periodic uprisings against the government, the 2007 protests represented a shift in forms of mobilization and publicity due to the use of the Internet and other communication technologies. In August 2007 NLD activists and members of other civil society organizations began street protests against rising living costs, prompted by the end of government subsidies on fuel. Buddhist monks joined the protests, leading to this uprising being termed the 'Saffron Revolution' due to the yellow colour of their robes. Military violence and oppression against the protesters generated further demonstrations. The government claims that 13 people were killed, while the UN Human Rights Council says there is evidence for 31 deaths and the pro-democracy groups argue there were hundreds.

After the initial repression international awareness of the events was prompted by images taken using mobile phones and then sent out of the country, often to Burmese exiles living in Thailand, Europe or the USA. This footage was often used by TV news broadcasters as they did not have permission to film the protests themselves. Images and information were also used by campaigning organizations outside Burma to highlight the repressive regime and to mobilize people to lobby their elected representatives and to engage in public protests in their own countries.

In response, the Burmese government shut down Internet connections in September 2007. They were able to do this because they controlled the only two Internet service providers in Burma. They also disabled international mobile phone coverage. This complete clampdown lasted for about two weeks, but government surveillance of Internet and mobile phone usage continues, with large numbers of websites that are seen as 'political' being blocked.

While the 'Saffron Revolution' did not succeed in overthrowing the government, it did highlight the repressive situation in Burma and provided significant evidence for campaigning groups outside the country. Mridul Chowdhury suggests that the circulation of images could have saved some lives as the government was wary that photographic evidence of brutality and killings would be used against the regime. However, Chowdhury also concludes that the regime's unwillingness to accept humanitarian aid following the devastation caused by Cyclone Nargis in May 2008 may be partly explained by their concerns about how modern media could be used as a tool against the government.

Source: adapted from Chowdhury (2008)

Networks of NGOs and community organizations are facilitated through the use of communication networks, and this technology can certainly help to overcome some of the problems of operating at a very small scale. As shown in Chapters 4, 5 and 6, while a grassroots approach can be very beneficial in terms of promoting participation and the involvement of local people, indigenous knowledge and appropriate technology in locally-defined development problems, there can be limits to the success of these projects in terms of scale. Being able to tap into larger-scale networks and be involved in umbrella organizations can begin to help overcome some of these limitations. Of course, this is not always the case, with numerous examples of umbrella organizations experiencing management problems, being too costly and bureaucratic.

Summary

- As globalization processes intensify, focusing development policies and theories purely at the national scale is difficult to justify.
- Neoliberal arguments support economic globalization, stressing the role of increased trade in poverty alleviation.

- Free trade is currently being promoted by the rules-based WTO, but it has been criticized for promoting the interests of the North over those of the South.
- The economic rise of countries such as China and India is changing the world's economic and political power structures.
- National governments and organizations have embraced the possibilities that aspects of globalization and transnationalism offer.

Discussion questions

1   What criticisms have been made of WTO policies and WTO activities?

2   How can globalization processes help alleviate poverty?

3   What role can the Internet play in development?

4   Do national governments still have a role in development policy in a globalizing world?

Further reading

Dicken, P. (2007) *Global Shift*, 5th edition, London: Sage. An excellent overview of economic globalization processes.

Murray, W.M. (2006) *Geographies of Globalization*, London: Routledge. Clear overview of geographical perspectives on globalization.

UNDP (2009) *Human Development Report 2009*, Basingstoke: Palgrave Macmillan (available at www.undp.org). The theme of this HDR was 'Overcoming human barriers: Human mobility and development'.

Williams, G., P. Meth and K. Willis (2009) *Geographies of Developing Areas: The Global South in a Changing World*, London: Routledge. Section 2 considers the Global South within economic, political, social and cultural globalization, focusing on how people and places in the Global South are both affected by and contribute to globalization.

Useful websites

www.enlacezapatista.ezln.org.mx   Zapatista network website in Spanish.

www.fairtrade.net   Fairtrade Labelling Organizations International. Overview of fair trade activities throughout the world.

www.fairtrade.org.uk   Fairtrade Foundation. Information about Fairtrade products in the UK and links to useful Fairtrade websites.

www.focac.org/eng/   Forum on China–Africa Cooperation.

www.forumsocialmundial.org.br/   World Social Forum homepage. The WSF was set up as a forum for NGOs, civil society organizations, social movements and other groups opposed to neoliberalism and imperialism.

http://cyber.law.harvard.edu/research/internetdemocracy   Internet and Democracy Project, Berkman Center for Internet and Society, Harvard University. Provides analysis of the role of the Internet in promoting democracy.

www.jubileeusa.org   Jubilee debt cancellation campaign in the USA.

www.kiva.org   Homepage of the Kiva organization that uses the Internet to link lenders to small-scale entrepreneurs.

www.resist.org.uk   Globalise Resistance. An organization that is against the growth of global corporate power.

www.tjm.org.uk   Trade Justice Movement.

http://viacampesina.org/en/   Via Campesina. An international movement of peasants, small-scale farmers and agricultural workers.

www.warwick.ac.uk/csgr/   Centre for the Study of Globalisation and Regionalistion, University of Warwick.

www.wto.org   World Trade Organization website. Includes description of WTO activities, why such an organization is beneficial and responds to 'common misconceptions' about the WTO.

8 Conclusions

- Summary of development theories
- Future of development theories and practices
- Cash transfers
- The BRICs
- Tobin Tax
- Post-development

In the previous seven chapters the diversity of development theories and practices has been very apparent. This diversity starts with the definitions of 'development'. While some theorists view economic growth and increases in economic wealth as the key definition of 'development', others consider 'development' to encompass ideas of greater autonomy and choice about how individuals live their lives (Table 8.1). In addition, 'development' can be seen as an end point to which particular societies aspire, and/or a process of change and what some have called 'progress'.

As well as the definitional differences, theories differ in who are the main actors called on to achieve 'development'. The main actors involved are governments or the state (at national or local levels), the market represented by private sector companies, and non-governmental organizations and civil society institutions. However, as outlined in Chapter 1, the role of individuals and communities is also important in both defining and achieving 'development', although the importance of their involvement will vary depending on the theory.

The scale of development theorizing also differs. Most development theories and many official development measures are based on the nation-state. However, in some cases, the focus on more local, grassroots activities and actors becomes key. In addition, the influence of global processes is implicated in a number of theories, most noticeably the structuralism and dependency theories of Latin

Table 8.1 Summary of development theories and approaches

Name	Main actors	Scale	Definition of development	Description
Classical economic theory	Private sector (the market)	National	Economic growth	Focus on market forces as the most efficient way of organizing economies
Classical Marxism	State	National	Economic growth, industrialization, urbanization, increased complexity of societies	State as key actor in organizing resource distribution and use
Keynesianism	State and market	National	Economic growth, in particular full employment	State intervention in the economy to help regions and groups that are disadvantaged
Modernization theory	State and market	National	Economic growth and increased complexity in social and economic organization	Eurocentric assumptions that all countries should follow the path of Northern nations
Structural approaches	State	National	Economic growth	National governments need to protect domestic production from global markets and competition because of global economic inequalities
Dependency theories	State	National	Economic growth	Economic disadvantage in the global periphery is a result of exploitation from the North; need to withdraw from global economic system
Neoliberalism	Private sector, NGOs and individuals	National and sub-national	Economic growth, liberal democracy	State involvement regarded as being detrimental to development; state should provide regulatory framework within which companies and NGOs can operate
Sustainable development	Depends on perspective	Depends on perspective	Protection of the natural environment	Diversity of approaches to sustainable development; some are very market-led and involve pricing nature, while others involve putting environmental protection at the heart of policy and reducing consumption
Ethnodevelopment	State and ethnic groups	National and sub-national	Recognition of ethnic diversity. Definitions may vary by ethnic group	Development decisions balance the requirements of different ethnic groups
Gender and development	Depends on perspective	National and sub-national	Moves towards greater gender equity	Approaches vary, but increasingly there is a focus on grassroots participation
Rights-based development	State, NGOs and individuals	Varies	Individuals and groups able to live fulfilled lives	Approaches vary from very small-scale awareness-raising activities to large-scale transnational campaigns
Post-development	Grassroots organizations and individuals	Very small-scale	A dangerous, Eurocentric concept which destroys local cultures and environments	Focus on grassroots activities, local-level participation

Plate 8.1 Colonial and capitalist influences, Hong Kong.
Credit: Katie Willis

American derivation. However, as outlined in Chapter 7, processes of globalization have increasingly challenged the ability of nation-states to manage and direct development within their own boundaries.

Finally, the location of development must be recognized. 'Development' is often regarded as something which only now happens in the Global South, as it has been achieved in the North. In fact, many development theories are based on the Northern experience which is then transposed to other parts of the world as the only correct way to develop. However, it is clear that regardless of definition, 'development' is an on-going process throughout the world, often with similar debates about appropriate policies. For example, processes of decentralization and participation are certainly not confined to the South (Jones 2000).

In terms of development practices, the post-Second World War development agenda epitomized by Truman's inaugural speech has led to a range of interventions by multilateral agencies and Northern governments in the South. While there have clearly been successes, the fact remains that the Millennium Development Goals (MDGs) are a reflection of the relative failure of development practices to provide even the most basic levels of food, shelter, healthcare and education for millions of the world's people. The likelihood of achieving the MDG targets by 2015 for all regions of the Global South is also very small despite concerted global efforts and the expenditure of billions of dollars (UNDP 2010b). For neoliberals, failure can be partly attributed to limited openness to trade, as well as the corruption and dependency that aid flows can provide (Collier 2007; Moyo 2010). For theorists from a Marxist perspective, the inequality of global power structures and the exploitative nature of capitalism will be blamed (Peet 2007).

The fate of Sub-Saharan Africa is of particular concern and in many ways 'development' has failed. As the UNDP (2010b) boldly states, in relation to MDG1 'Eradicate extreme poverty and hunger':

> The goal of cutting in half the proportion of people in the developing world living on less than $1 a day by 2015 remains within reach. However, this achievement will be due largely to extraordinary economic success in most of Asia. In contrast, previous estimates suggest that little progress was made in reducing extreme poverty in sub-Saharan Africa.

I do not want to reinforce the common representations of the whole continent as poverty-stricken and in need of assistance, and it must be remembered that the MDGs are an example of a top-down development definition, but millions of African people are living in conditions where their life choices are highly limited. For example, an estimated 30 per cent of the Sub-Saharan population was undernourished in the period 2004–6, with a figure of 57 per cent for Central African states (FAO 2010).

International concerns about the region have led to a number of high-profile initiatives such as the 'New Partnership for Africa's Development' (NEPAD), which was established in 2001 and promotes partnership between African governments, donors and African citizens. It has adopted a largely neoliberal approach to national and regional economic strategy (Owusu 2003). Such initiatives may have a positive impact, but only if they can help create global conditions that are more conducive to the enhancement of human freedoms; in particular, more equitable participation in global governance institutions and fairer trade regimes.

Future of development theories and practices

Development theory comes out of contemporary situations and problems, but also feeds into policy-making processes. At the end of the first decade of the twenty-first century, neoliberal approaches to development dominate the political agenda in both multilateral organizations such as the World Bank and the World Trade Organization, and also at the level of many national governments. This is despite the criticisms levelled at such approaches from a range of sources (see Chapters 2 and 7), as well as the perceived failures of the neoliberal models that led to the global economic crisis.

However, neoliberalism is not a homogenous process and it can be interpreted and implemented in different ways (Willis *et al.* 2008). Following the very harsh implementation of structural adjustment policies in the 1980s and 1990s, there has been a growing recognition by IFIs and Northern governments that policies need to ensure that the poorest sectors of society are protected. This has included more PSRPs, although these have not always had the desired effect (see Chapter 2). A policy that has grown in popularity has been cash transfers.

As the name suggests, 'cash transfers' involve giving identified poor or marginalized individuals or households money (Hanlon *et al.* 2010). This usually has some conditions attached, so often they are termed 'conditional cash transfers'. They have been most widely adopted in Latin America, with Brazil's 'Bolsa Familia' and Mexico's 'Oportunidades' schemes being among the most well-known. Under the Oportunidades programme, which had an estimated 25 million beneficiaries in 2005, money is given to women if their children go to school and attend the health centre for check-ups (Molyneux 2006). Greater amounts are given for girls attending school to encourage gender equality in educational opportunities. Providing money directly to households in this way has resulted in improved health, nutritional status and educational levels (Hanlon *et al.* 2010).

While the money in these conditional cash transfer schemes is usually channelled through government routes, neoliberal institutions, such as the World Bank, have been attracted to these schemes as they are interpreted as giving individuals and households greater freedom from state control. They are also seen to be successful in reducing indicators of poverty. However, as with the discussion of participation in Chapter 4, opponents of neoliberalism have also been attracted to cash transfer programmes because they are seen as potentially empowering. Maxine Molyneux (2006) suggests that some of the conditions placed on women within the Oportunidades programme are viewed as an additional time and energy burden (see Chapter 5 for similar discussions regarding SAPs), but that the provision of cash to the household can be very beneficial overall.

The rise of the BRICs

One of the major shifts in development research and policy concerns since the first issue of this book has been the challenges that result from the rise of Brazil, Russia, India and China. This is likely to continue in the future, and will have an impact on development, both within these particular countries and also globally.

There are three main areas that can be highlighted at this point. The first relates to economic flows, particularly foreign direct investment and trade. The rapidly growing importance of China as a global trading partner was discussed in Chapter 7, as was the importance of Chinese and Indian FDI in parts of Africa. However, as well as the current and potential importance of these FDI flows on African economies and possible job creation, Raphael Kaplinsky (2008) highlights other potential economic impacts of the 'Asian Drivers' of China and India. This includes both undermining domestic industrial production due to cheaper Chinese imports, but also possibly usurping African clothing, textile and furniture sales to other parts of the world. For Kaplinsky, protectionism either by individual African economies or as regional trading blocs, is one possible response to these perceived threats.

In Chapter 2, the growth of the non-DAC donors in global aid flows was discussed. The BRIC economies, as well as the oil-rich Gulf States and the Latin American countries involved in ALBA, are all examples of participants in forms of international aid that fall outside the North–South patterns that were dominant in the past. Because of this, they can frame their overseas assistance in terms of collaboration and a partnership of equals (although in reality this is often not the case). In relation to Africa, China provides (among other things) loans and grants for infrastructure construction, as well as training, technical assistance and debt relief. In contrast, India has tended to focus on training and technical assistance, rather than monetary support (McCormick 2008). The rise of these non-DAC donors will require the 'traditional' donors of the Global North and multilateral institutions to reconsider the support they provide when recipient countries have greater choice.

Finally, the rising economic might of a number of Southern countries and Russia also has implications for global governance institutions and the balance of power. As was outlined in Chapter 7, international institutions supposedly operating in a neutral way have sometimes been accused of following a Northern agenda. The failures of successive trade discussions within the WTO, as well as the power struggles evident at the Copenhagen Climate Conference (Chapter 6) indicate that future development policy and international governance structures may reflect these shifts in global power. However, what is harder to assess, is whether this will lead to a better and more inclusive global system for the world's poorest people, or whether it will just mean different faces around the main table.

Tobin Tax

As outlined at the start of this chapter, when considering the future of 'development' theories and policies throughout the world, the over-arching influence of neoliberalism and market-centred theories is never far away. Government responses to the global economic crisis have reflected some disquiet with the status quo, not least in relation to bank regulation. There have also been attempts to boost national economies through neo-Keynesian policies (see Chapter 2), but the commitment to overall free market systems remains in most countries (although see the discussion of Latin American socialist countries in Chapter 3).

However, in the aftermath of the crisis some national leaders have looked to a seemingly radical idea: the Tobin Tax. This is an idea proposed by the economist James Tobin in the mid-1970s. The main idea behind the tax is to reduce the amount of short-term capital flow in and out of countries by taxing financial transactions, and to generate revenue to be spent on poverty alleviation throughout the world. With the increasing liberalization of global money markets, financial transactions have increased enormously. As the Asian financial crisis (Chapter 2) demonstrated, these transactions are often based on very short-term investments, allowing money to be withdrawn and moved elsewhere in times of uncertainty. The global economic crisis of 2007 onwards also highlighted the fragility of financial markets due to complex financial products and the overstretching of financial institutions in search of greater profit.

A tax of this type could also be used to fund a range of activities. It is estimated that taxing existing financial transactions at a rate of 0.05 per cent could raise US$250 billion per annum (War on Want 2010). The Millennium Development Targets would certainly be more achievable if this level of funding was available. While there are clearly issues about how the tax would be collected, and there are fears of tax evasion (Raffer 1998), the main obstacle to its implementation is political will. Tobin wrote about his global taxation idea to the 1994 *Human Development Report* (Tobin 1994), but it had never been embraced as a serious proposal by Northern governments or IFIs before the global crisis. Kunibert Raffer claims that 'the main reason for the proposal's unpopularity is that Keynesian ideas, more government influence and raising money for international projects runs counter to presently ruling neoliberalism' (1998: 529). While ideas of a 'transaction tax' have been discussed at G20 meetings and the IMF, it remains very far from being adopted due to continued opposition, not least from the banking sector (Inman 2010).

Post-development

The Tobin Tax and other initiatives, such as debt relief and opening up Northern markets to Southern exports, could be criticized as still operating within a capitalist system with decisions being made at a national or supra-national level. As we have seen throughout the book, post-development theorists have argued that 'development' should be focused on what local communities want and should not be a response to a Northern-imposed model of what is a correct form of development.

The post-development approach has been very important in highlighting the ways in which 'development' as a concept is always a product of a particular set of power relations at any one time. It has also been crucial in stressing the importance of discourse, or the language that we use to describe particular processes (Sachs 1992).

However, in other senses, post-development has been greatly criticized. First, many have argued that the 'development' which post-development theorists criticize is a caricature of the diversity of development approaches today (Corbridge 1998; Rigg 2003; Simon 1998). Escobar's highly influential study (see Chapter 1) is based largely on development policies in Colombia in the 1950s and 1960s. Rather than recognizing the diversity and dynamism of development theorizing and practice, much post-development analysis has drawn on a stereotypical image of how development has been defined and implemented. Second, post-development approaches have been accused of making substantial criticisms of current policies and theories without providing details of possible alternatives, other than focusing on grassroots communities. Finally, while 'development' in many guises has not brought all the benefits it was purported to provide, there have been significant material improvements in life expectancy, health levels and education for some populations and regions (Rigg 2003).

James Sidaway (2008) in his review of post-development approaches, recognizes these criticisms and others. However, he does conclude by stressing the contributions which the range of theorists under the 'post-development' heading have made; most explicitly, raising the issue of the way in which 'development' has often been defined and presented with no consideration of the social, political and economic context of these formulations. This book has highlighted how the definitions of development, development theories and policies have been formed over time and at a range of scales. Given the dynamic nature of twenty-first-century societies and economies, this diversity will continue as governments, NGOs and local communities continue to find ways to improve people's lives.

Neoliberalism remains the broad theoretical context which shapes so much of international development policy today despite the devastating effects of SAPs and related policies on many communities, households and individuals, as well as national economies. Since the late 1990s, IFIs have addressed social development and poverty-alleviation more explicitly, but this change of focus has not meant a shift in the underlying philosophies about the route to development. While the increased importance of grassroots initiatives and a focus on rights could be interpreted as positive steps towards people-centred processes and definitions of 'development', far too often these trends have been shaped by a continued faith in the market as the key actor in development. Given current global inequalities in economic and political power and relationships of dependence, the scope for autonomous development decisions by the peoples of the South remains a distant dream. This does not meant that positive changes cannot come about, rather that the scope of 'people-centred development' will remain limited by broader structural factors, particularly at a global scale.

Summary

- Development theories differ by definitions of 'development', key actors and approaches.
- Over time, what is regarded as 'development' has become more complex and diverse.
- Neoliberalism has become the key theory informing global development policy, but it can be implemented in a number of ways at a range of scales.
- The rise of emerging powers such as China and India is creating new opportunities and challenges for development theory and practice.
- The Tobin Tax has been suggested as a way of both regulating global financial flows and generating finance to fund poverty-alleviation strategies.
- Post-development stresses the socially-constructed nature of development, but has been criticized for its tendency to homogenize 'development' in its critiques.

Discussion questions

1   What are cash transfers and what contribution can they make to poverty alleviation in the Global South?

2   What are the advantages and disadvantages of implementing the Tobin Tax?

3   What are the impacts of the rise of Chinese and Indian investment, trade and aid flows to Africa?

4   What contributions have post-development approaches made to our understandings of 'development'?

Further reading

Corbridge, Stuart (1998) '"Beneath the pavement only soil": the poverty of post development', *Journal of Development Studies* 34 (6): 138–48. A critique of post-development approaches to development.

Hanlon, J., A. Barrientos and D. Hulme (2010) *Just Give Money to the Poor: The Development Revolution from the Global South*, Sterling, VA: Kumarian Press. A clear and engaging discussion of cash transfer programmes.

Jones, P.S. (2000) 'Why is it alright to do development "over there" but not "here"? Changing vocabularies and common strategies of inclusion across "First" and "Third" Worlds', *Area* 32 (2): 237–41. Very accessible discussion of how 'development' is portrayed as occurring only in the South.

Owusu, F. (2003) 'Pragmatism and the gradual shift from dependency to neoliberalism: The World Bank, African leaders and development policy in Africa', *World Development* 31 (10): 1655–72. Clear overview of changing approaches to African development, contrasting the World Bank neoliberal perspectives with African explanations.

Review of African Political Economy (2008) 'Special issue on China and India in Africa', *Review of African Political Economy*, 34 (115). A collection of full-length papers and short briefings on India and China's influence in Africa.

Useful websites

www.bwpi.manchester.ac.uk/ Brooks World Poverty Institute, University of Manchester. Provides access to publications on cash transfers.

www.nepad.org NEPAD website.

www.waronwant.org/campaigns/financial-crisis/the-robin-hood-tax Useful website on the Tobin Tax run by War on Want. They have termed it the 'Robin Hood Tax' in the aftermath of the global economic crisis.

Bibliography

Aalbers, M. (2009) 'Geographies of the financial crisis', *Area* 41 (1): 34–42.

Adger, N., S. Huq, K. Brown, D. Conway and M. Hulme (2003) 'Adaptation to climate change in the developing world', *Progress in Development Studies* 3 (3): 179–95.

African Union (2010) 'African Union homepage', http://www.africa.union.org (accessed 28 July 2010).

Afshar, H. (1985) 'Women, the state and ideology in Iran', *Third World Quarterly* 7 (2): 256–78.

AGRA (2010) 'Alliance for a Green Revolution in Africa', www.agra-alliance.org/ (accessed 29 July 2010).

Ahmad, M.M. (2006) 'The 'partnership' between international NGOs (non-governmental organisations) and local NGOs in Bangladesh', *Journal of International Development* 18 (5): 629–38.

Alexander, R.J. (1999) *International Maoism in the Developing World*, London: Praeger.

Alkire, S. and M.E. Santos (2010) 'Multidimensional Poverty Index', *OHPI Research Brief, July 2010* (available at www.ophi.org.uk/wp-content/uploads/OPHI-MPI-Brief.pdf).

Allen, J. (1995) 'Global worlds', in J. Allen and D. Massey (eds) *Geographical Worlds*, Oxford: Oxford University Press, pp. 105–42.

Amin, S. (1974) *Accumulation on a World Scale: A Critique of the Theory of Underdevelopment*, New York: Monthly Review Press.

Amsden, A.H. (1994) 'Why isn't the whole world experimenting with the East Asian model to develop?: review of *The East Asian Miracle*', *World Development* 22 (4): 627–33.

Anderson, B. (2001) 'Different roots in common ground: transnationalism and migrant domestic workers in London', *Journal of Ethnic and Migration Studies* 27 (4): 673–83.

APEC (2010) 'About APEC', www.apec.org./apec/about_apec.html (accessed 28 July 2010).

ASEAN (2010) 'Overview: Association of Southeast Asian Nations', http://www.asean.org/about_ASEAN.html (accessed 28 July 2010).

Ashwin, S. (2000) 'Introduction: gender, state and society in Soviet and post-Soviet Russia', in S. Ashwin (ed.) *Gender, State and Society in Soviet and Post-Soviet Russia*, London: Routledge, pp. 1–29.

Ayers, J. and D. Dodman (2010) 'Climate change adaptation and development I: the state of the debate', *Progress in Development Studies* 10 (2): 161–8.

Balassa, B. (1971) 'Trade policies in developing countries', *American Economic Review* 61 (May): 178–87.

Balassa, B. (1981) *The Newly Industrializing Countries in the World Economy*, Oxford: Pergamon.

Baral, N. and M.J. Stern (2009) 'Looking back and looking ahead: local empowerment and governance in the Annapurna Conservation Area, Nepal', *Environmental Conservation* 37 (1): 54–63.

Baran, P. (1960) *The Political Economy of Growth*, New York: Monthly Review Press.

Baran, P. and P. Sweezy (1968) *Monopoly Capital*, Harmondsworth: Penguin.

Barefoot College (2010) 'Barefoot College', http://www.barefootcollege.org (accessed 1 August 2010).

Barr, M.D. and Z. Skrbis (2008) *Constructing Singapore: Elitism, Ethnicity and the Nation-Building Project*, Copenhagen: NIAS Press.

Barrow, C.J. (1995) *Developing the Environment: Problems and Management*, Harlow: Longman.

Barrow, C.J. (2006) *Environmental Management for Sustainable Development*, 2nd edition, London: Routledge.

Bauer, P. (1972) *Dissent on Development*, Boston: Harvard University Press.

Beales, S. (2000) 'Why we should invest in older women and men: the experience of HelpAge International', *Gender and Development* 8 (2): 9–18.

Bebbington, A. (1997) 'Social capital and rural intensification: local organisations and islands of sustainability in the rural Andes', *Geographical Journal* 163 (2): 189–97.

Bebbington, A. (2007) 'Social capital and development studies II: can Bourdieu travel to policy?', *Progress in Development Studies* 7 (2): 155–62.

Bebbington, A. (2008) 'Social capital and development', in V. Desai and R. Potter (eds) *The Companion to Development Studies*, 2nd edition, London: Hodder Education, pp. 132–6.

Bello, W., S. Cunningham and Li K.P. (1998) *A Siamese Tragedy: Development and Disintegration in Modern Thailand*, London: Zed Books.

Benería, L. and G. Sen (1981) 'Accumulation, reproduction and women's role in economic development: Boserup revisited', *Signs* 8 (2): 279–98.

Bernstein, H. (2000) 'Colonialism, capitalism, development', in T. Allen and A. Thomas (eds) *Poverty and Development into the 21st Century*, Oxford: Oxford University Press, pp. 241–70.

Binns, T. (2008) 'Dualistic and unilinear concepts of development', in V. Desai and R.B. Potter (eds) *The Companion to Development Studies*, 2nd edition, London: Hodder Education, 81–6.

Blaut, J. (1993) *The Colonizer's Model of the World: Geographic Diffusionism and Eurocentric History*, New York: Guilford Press.

Blowfield, M. (1999) 'Ethical trade: a review of developments and issues', *Third World Quarterly* 20 (4): 753–70.

Booth, D. (1985) 'Marxism and development sociology: interpreting the impasse', *World Development* 13 (7): 761–87.

Boserup, E. (1965) *The Conditions of Agricultural Growth: The Economics of Agrarian Change under Population Pressure*, London: Allen and Unwin.

Boserup, E. (1989) *Woman's Role in Economic Development*, London: Earthscan [originally published in 1970].

Boyd, D. (1987) 'The impact of adjustment policies on vulnerable groups: the case of Jamaica, 1973–1985', in G.A. Cornia, R. Jolly and F. Stewart (eds) *Adjustment with a Human Face: Ten Case Studies*, Oxford: Clarendon Press, pp. 126–55.

Bradley, T. (2009) 'A call for clarification and critical analysis of the work of faith-based development organisations (FBDO)', *Progress in Development Studies* 9 (2): 101–14.

Bradshaw, M. and A. Stenning (eds) (2004) *East Central Europe and the Former Soviet Union*, London: Pearson Prentice Hall.

Bradshaw, S. and B. Linneker (2003) 'Civil society response to poverty reduction strategies in Nicaragua', *Progress in Development Studies* 3 (2): 147–58.

Brandt Commission (1980) *North–South: A Programme for Survival*, London: Pan.

Branigan, T. (2009) 'Unemployment forces Chinese migrants back to the countryside', *The Guardian*, 17 May 2009.

Brohman, J. (1996) *Popular Development: Rethinking the Theory and Practice of Development*, Oxford: Blackwell.

Bujra, J. (2000) 'Diversity in pre-capitalist societies', in T. Allen and A. Thomas (eds) *Poverty and Development into the 21st Century*, Oxford: Oxford University Press, pp. 219–40.

Bulmer, M. and D.P. Warwick (eds) (1993) *Social Research in Developing Countries: Surveys and Censuses in the Third World*, London: UCL Press.

Bulmer-Thomas, V. (1988) 'The Central American Common Market', in A.M. El-Agraa (ed.) *International Economic Integration*, 2nd edition, Basingstoke: Macmillan, pp. 253–77.

Bulmer-Thomas, V. (1998) 'The Central American Common Market: from closed to open regionalism', *World Development* 26 (2): 313–22.

Bunun Cultural and Educational Foundation (2010) 'Bunun Cultural and Education Foundation homepage', www.bunun.org.tw/ (accessed 7 August 2010).

Butler, R. (1998) 'Sustainable tourism – looking backwards in order to progress?', in C.M. Hall and A.A. Lew (eds) *Sustainable Tourism: A Geographical Perspective*, Harlow: Longman, pp. 25–34.

CARICOM (2010) 'The Caribbean Community', www.caricom.org/jsp/ community/community_index.jsp?menu=community (accessed 19 July 2010).

Carson, R. (1962) *Silent Spring*, St Louis: Houghton Mifflin.

Cassen, R. and associates (1994) *Does Aid Work?*, 2nd edition, Oxford: Clarendon Press.

Castells, M. (2004) *The Power of Identity*, 2nd edition, Oxford: Blackwell.

Castles, S. and M.J. Miller (2009) *The Age of Migration: International Population Movements in the Modern World*, 4th edition, Basingstoke: Palgrave Macmillan.

Chambers, R. (1997) *Whose Reality Counts? Putting the First Last*, London: Intermediate Technology Books.

Chant, S. (1994) 'Women's work and household survival strategies in Mexico, 1982–1992: past trends, current tendencies and future research', *Bulletin of Latin American Research* 13 (2): 203–33.

Chant, S. with N. Craske (2003) *Gender in Latin America*, London: Latin American Bureau.

Chant, S. and M. Gutmann (2002) '"Men-streaming" gender? Questions for gender and development policy in the twenty-first century', *Progress in Development Studies* 2 (4): 269–82.

Chant, S. and C. McIlwaine (2009) *Geographies of Development in the 21st Century*, Cheltenham: Edward Elgar.

Chenery, H. (1989) 'Foreign aid', in J. Eatwell, M. Milgate and P. Newman (eds) *The New Palgrave: Economic Development*, London: Macmillan, pp. 137–44.

Chetley, A. (1995) 'Paying for health: new lessons from China', *IDS Policy Briefing* Issue 4.

Chowdhury, M. (2008) 'The role of the internet in Burma's Saffron Revolution', Berkman Center Research Publication 2008, Cambridge Mass: Berkman Center for Internet & Society at Harvard University (available at: http://cyber.law.harvard.edu/sites/cyber.law.harvard.edu/files/ Chowdhury_Role_of_the_Internet_in_Burmas_Saffron_Revolution. pdf_0.pdf).

Clarke, C. (2002) 'The Latin American structuralists', in V. Desai and R.B. Potter (eds) *The Companion to Development Studies*, London: Arnold, pp. 92–6.

Cohen, R. and S.M. Rai (2000) 'Global social movements: towards a cosmopolitan politics', in R. Cohen and S.M. Rai (eds) *Global Social Movements*, London: Athlone Press, pp. 1–17.

Collier, P. (2007) *The Bottom Billion*, Oxford: Oxford University Press.

Cooke, B. and U. Kothari (eds) (2001) *Participation: The New Tyranny?*, London: Zed Books.

Corbridge, S. (1998) 'Beneath the pavement only soil: the poverty of post-development', *Journal of Development Studies* 34 (6): 138–48.

Cornia, G.A., R. Jolly and F. Stewart (1987) *Adjustment with a Human Face: Protecting the Vulnerable and Promoting Growth*, Oxford: Clarendon Press.

Cornwall, A., S. Corrêa and S. Jolly (2008) 'Development with a body: making connections between sexuality, human rights and development', in A. Cornwall, S. S. Corrêa and S. Jolly (eds) *Development with a Body: Sexuality, Human Rights & Development*, London: Zed Books, pp. 1–21.

Dalla Costa, M. (1995) *Paying the Price: Women and the Politics of International Economic Strategy*, London: Zed Books.

Davies, M. (2008) 'China's developmental model comes to Africa', *Review of African Political Economy* 35 (115): 134–7.

Davies, R. (2010) 'Development challenges for a resurgent African diaspora', *Progress in Development Studies* 10 (2): 131–44.

Davies, R.W. (1998) *Soviet Economic Development from Lenin to Khrushchev*, Cambridge: Cambridge University Press.

De Haas, H. (2006) 'Migration, remittances and regional development in Southern Morocco', *Geoforum* 37 (4): 565–80.

de Waal, A. (2002) 'What's new in the "New Partnership for Africa's Development"?', *International Affairs* 78 (23): 463–75.

Department for International Development (DFID) (1997) *Eliminating World Poverty: A Challenge for the Twenty-first Century*, White Paper on International Development, London: HMSO (available on http://www.dfid.gov.uk).

Department for International Development (DFID) (2000a) *Eliminating World Poverty: Making Globalisation Work for the Poor*, White Paper on International Development, London: HMSO (available on http://www.dfid.gov.uk).

Department for International Development (DFID) (2000b) *Disability, Poverty and Development*, London: DFID (available on www.inclusive-development.org/docsen/DFIDdisabilityPovertyDev.pdf).

Department for International Development (DFID) (2009) 'Statistics on international development, 2009' (available at www.dfid.gov.uk/About-DFID/Finance-and-performance/Aid-Statistics/Statistics-on-International-Development-2009/).

Desai, V. and M. Tye (2009) 'Critically understanding Asian perspectives on ageing', *Third World Quarterly* 30 (5): 1007–25.

Development Goals (2010) 'Millennium development goals'. http://www.developmentgoals.org/About_the_goals.htm (accessed 14 July 2010).

Diacon, D. (1998) *Housing the Homeless in Ecuador: Affordable Housing for the Poorest of the Poor*, Coalville: Building and Social Housing Foundation.

Dicken, P. (2003) *Global Shift*, 4th edition, London: Sage.

Dicken, P. (2004) 'Geographers and "globalization": (yet) another missed boat?', *Transactions of the Institute of British Geographers* 29 (1): 5–26.

Dicken, P. (2007) *Global Shift: Mapping the Changing Contours of the World Economy*, 5th edition, London: Sage.

Divine Chocolate (2010) 'Divine Chocolate homepage' http://www.divinechocolate.com (accessed 31 July 2010).

Dixon, C. (1999) 'The Pacific Asian challenge to neoliberalism', in D. Simon and A. Närman (eds) *Development as Theory and Practice*, Harlow: Longman, pp. 205–29.

Dossani, R. and M. Kenney (2007) 'The next wave of globalization: relocating service provision to India', *World Development* 35 (5): 772–91.

Drakakis-Smith, D. (1995) 'Third World cities: sustainable urban development I', *Urban Studies* 32 (4–5): 659–77.

Drakakis-Smith, D. (1996) 'Third World cities: sustainable urban development II', *Urban Studies* 33 (4–5): 673–701.

Drakakis-Smith, D. (1997) 'Third World cities: sustainable urban development III', *Urban Studies* 34 (5–6): 797–823.

Duffield, M. (2007) *Development, Security and Unending War: Governing the World of Peoples*, Cambridge: Polity.

Durkheim, E. (1966) *The Division of Labour in Society*, trans. George Simpson, New York: Free Press [originally published in French in 1893].

Economist, The (2010a) 'Speak softly and carry a blank cheque', *The Economist*, 17 July 2010: 42–3.

Economist, The (2010b) 'The world economy calls', *The Economist*, 25 March 2010: 70.

Economist, The (2003) 'Would you like your class war shaken or stirred?', *The Economist* 6 September: 46–7.

Edwards, M. (1996) 'New approaches to children and development: introduction and overview', *Journal of International Development* 8 (6): 813–27.

Edwards, M. and D. Hulme (1995) 'NGO performance and accountability: introduction and overview' in M. Edwards and D. Hulme (eds) *Non-Governmental Organisations – Performance and Accountability: Beyond the Magic Bullet*, London: Earthscan, pp. 3–16.

Elliott, J.A. (2006) *An Introduction to Sustainable Development*, 3rd edition, London: Routledge.

Elson, D. (1995) 'Male bias in macro-economics: the case of structural adjustment', in D. Elson (ed.) *Male Bias in the Development Process*, 2nd edition, Manchester: Manchester University Press, pp. 164–90.

Engels, F. (1940) *The Origin of the Family, Private Property and the State*, London: Lawrence and Wishart Ltd [originally published in German in 1884].

Engels, F. (1984) *The Condition of the Working Class in England*, London: Granada [originally published in German in 1845].

Escobar, A. (1995) *Encountering Development: The Making and Unmaking of the Third World*, Princeton: Princeton University Press.

Esteva, G. and M.S. Prakash (1997) 'From global thinking to local thinking', in M. Rahnema with V. Bawtree (eds) *The Post-Development Reader*, London: Zed Books, pp. 277–89.

Europa (2010) 'Europa: Official website of the European Union'
www.europa.org (accessed 29 July 2010).

Fairris, D. and M. Reich (2005) 'The impact of living wage policies:
introduction to the Special Issue', *Industrial Relations* 44 (1): 1–13.

Fairtrade Foundation (2010) 'Facts and figures on Fairtrade' www.fairtrade.org.
uk/what_is_fairtrade/facts_and_figures.aspx (accessed 28 July 2010).

Fairtrade Labelling Organizations International (FLO) (2010) 'FLO homepage',
http://www.fairtrade.net (accessed 31 July 2010).

Fanon, F. (1986) *Black Skin, White Masks*, London: Pluto Press [originally
published in French in 1952 by Editions du Seuil].

Food and Agriculture Organization (FAO) (2010) 'Food security statistics',
www.fao.org/economic/ess/food-security-statistics/en/ (accessed 10 August
2010).

Farrington, J. and A. Bebbington (1993) *Reluctant Partners?*, London:
Routledge.

Fine, B. (2008) 'Social capital in wonderland: the World Bank behind the
looking glass', *Progress in Development Studies* 8 (3): 261–9.

Fine, B. (2010) *Theories of Social Capital: Researchers Behaving Badly*,
London: Pluto Press.

Forbes, D. and N. Thrift (1987) 'Introduction', in D. Forbes and N. Thrift (eds)
The Socialist Third World: Urban Development and Territorial Planning,
Oxford: Blackwell, pp. 1–26.

Francis, P. (2001) 'Participatory development at the World Bank: the primacy
of process', in B. Cooke and U. Kothari (eds) *Participation: The New
Tyranny?*, London: Zed Books, pp. 72–87.

Frank, A.G. (1967) *Capitalism and Underdevelopment in Latin America*,
London: Monthly Review Press.

Franke, R. (2002) 'Kerala fact sheet 2002' (available at http://chss.montclair.
edu/anthro/FactSheet.htm).

Frayne, B. (2010) 'Pathways of food: migration and food security in Southern
African cities', *International Development Planning Review* 32: (3–4): 291–310.

Friedmann, J. (1992a) *Empowerment: The Politics of Alternative Development*,
Oxford: Blackwell.

Friedmann, J. (1992b) 'The end of the Third World', *Third World Planning
Review* 14 (3): iii–vii.

Fukuyama, F. (1989) 'The end of history?', *The National Interest* 16: 3–18.

Furtado, C. (1976) *Economic Development in Latin America*, 2nd edition,
Cambridge: Cambridge University Press.

Gandy, M. (2006) 'Planning, anti-planning and the infrastructure crisis facing
Metropolitan Lagos', *Urban Studies* 43 (2): 371–96.

Garikipati, S. and W. Olsen (2008) 'The role of agency in development
planning and the development process', *International Development Planning
Review* 30 (4), 327–38.

Garnaut, R. (1998) 'The East Asian crisis', in R.H. McLeod and R. Garnaut
(eds) *East Asia in Crisis: From Being A Miracle to Needing One?*, London:
Routledge, pp. 3–27.

Gibb, R. (2004) 'International and regional trade in eastern and southern Africa', in D. Potts and T. Bowyer-Bower (eds) *Eastern and Southern Africa: Development Challenges in a Volatile Region*, London: Pearson, pp. 295–327.

Gibson-Graham, J.K. (2006) *A Postcapitalist Politics*, Minneapolis: University of Minnesota Press.

Glassman, J. (2001) 'From Seattle (and Ubon) to Bangkok: the scales of resistance to corporate globalization', *Environment and Planning D: Society and Space*, 19: 513–33.

Glick Schiller, N., L. Basch and C. Szanton Blanc (1992) 'Transnationalism: a new analytic framework for understanding migration', in N. Glick Schiller, L. Basch and C. Szanton Blanc (eds) *Towards a Transnational Perspective on Migration: Race, Class, Ethnicity and Nationalism Reconsidered*, New York: New York Academy of Sciences, pp. 1–24.

Goodman, D. and M.J. Watts (eds) (1997) *Globalising Food: Agrarian Questions and Global Restructuring*, London: Routledge.

Govmonitor (2009) 'Australia highlights Overseas Development Assistance partnership with NGOs' (available at http://thegovmonitor.com/health/australia-highlights-overseas-development-assistance-partnership-with-ngos-18820.html).

Grameen Bank (2010) 'Grameen Bank homepage' www.grameen-info.org/ (accessed 3 August 2010).

Gready, P. and J. Ensor (2005) 'Introduction', in P. Gready and J. Ensor (eds) *Reinventing Development: Translating Rights-Based Approaches from Theory into Practice*, London: Zed Books, pp. 1–44.

Green, A. and A. Matthias (1995) 'NGOs – A policy panacea for the next millennium', *Journal of International Development* 7 (3): 565–73.

Green, D. (2008) *From Poverty to Power: How Active Citizens and Effective States can Change the World*, Oxford: Oxfam.

Grootaert, C. (1998) 'Social capital: the missing link?', *Social Capital Initiative Working Paper No. 3*, Washington DC: World Bank (available at: http://siteresources.worldbank.org/INTSOCIALCAPITAL/Resources/Social-Capital-Initiative-Working-Paper-Series/SCI-WPS-03.pdf).

Grootaert, C., D. Narayan, V.N. Jones and M. Woolcock (2004) 'Measuring social capital: an integrated questionnaire', *World Bank Working Paper No. 18*, Washington DC: World Bank.

Gross National Happiness (2010) 'Gross National Happiness homepage', www.grossnationalhappiness.com/ (accessed 9 August 2010).

Grugel, J. (1995) *Politics and Development in the Caribbean Basin*, Basingstoke: Macmillan.

Guardian, The (2010) 'US bailouts prevented 1930s-style Great Depression say economists', *The Guardian*, 28 July 2010 (available at: www.guardian.co.uk).

Guardian, The (2010) 'US economy shows signs of slowdown as consumer spending falters', *The Guardian*, 30 July 2010 (available at: www.guardian.co.uk).

Gwynne, R. (1996) 'Industrialisation and urbanisation', in D. Preston (ed.) *Latin American Development: Geographical Perspectives*, 2nd edition, Harlow: Longman, 216–45.

Hall, C.M. and A.A. Lew (eds) (1998) *Sustainable Tourism: A Geographical Perspective*, Harlow: Longman.

Handa, S. and D. King (1997) 'Structural adjustment policies, income distribution and poverty: a review of the Jamaican experience', *World Development* 25 (6): 915–30.

Handa, S. and D. King (2003) 'Adjustment with a human face? Evidence from Jamaica', *World Development* 31 (7): 1125–45.

Handicap International (2010) 'Handicap International homepage', www.handicap-international.org.uk (accessed 9 August 2010).

Hanlon, J. (1991) *Mozambique: Who Calls the Shots?*, London: James Currey.

Hanlon, J., A. Barrientos and D. Hulme (2010) *Just Give Money to the Poor: The Development Revolution from the Global South*, Sterling, VA: Kumarian Press.

Happy Planet Index (2010) 'Happy Planet Index homepage', www.happyplanetindex.org/ (accessed 9 August 2010).

Hargreaves, J.R., L.A. Morison, J.S.S. Gear, M.K. Makhubele, J.D.H. Porter, J. Busza, C. Watts, J.C. Kim and P.M. Pronyk (2007) '"Hearing the voices of the poor": assigning poverty lines on the basis of local perceptions of poverty. A quantitative analysis of qualitative data from participatory wealth ranking in rural South Africa', *World Development* 35 (2): 212–29.

Harrison, G. (2005) 'The World Bank, governance and theories of political action in Africa, *British Journal of Politics and International Relations* 7 (2): 240–70.

Hart, G. (2001) 'Development critiques in the 1990s: *culs de sac* and promising paths', *Progress in Human Geography* 24 (4): 649–58.

Harvey, D. (2007) *A Brief History of Neoliberalism*, Oxford: Oxford University Press.

Harvey, D. (2010) *The Enigma of Capital and the Crises of Capitalism*, London: Profile Books.

Hashemi, S. (1995) 'NGO accountability in Bangladesh: beneficiaries, donors and the state', in M. Edwards and D. Hulme (eds) *Non-Governmental Organisations – Performance and Accountability: Beyond the Magic Bullet*, London: Earthscan, pp. 103–10.

Hashemi, S.J., S.R. Schuler and A.P. Riley (1996) 'Rural credit programs and women's empowerment in Bangladesh', *World Development* 24 (4): 635–63.

Havemann, P. (2000) 'Enmeshed in the web?: indigenous peoples' rights in the network society', in R. Cohen and S.M. Rai (eds) *Global Social Movements*, London: Athlone Press, pp. 18–32.

Held, D., A. McGrew, D. Goldblatt and J. Perraton (1999) *Global Transformations: Politics, Economics and Culture*, Cambridge: Polity.

HelpAge International (2010) 'HelpAge homepage', http://www.helpage.org (accessed 1 August 2010).

Hettne, B. (1995) *Development Theory and the Three Worlds*, 2nd edition, Harlow: Longman.

Hettne, B. (1996) 'Ethnicity and development: an elusive relationship', in D. Dwyer and D. Drakakis-Smith (eds) *Ethnicity and Development: Geographical Perspectives*, London: John Wiley and Sons Ltd, pp. 16–44.

Hickey, S. and G. Mohan (eds) (2004) *Participation: From Tyranny to Transformation*, London: Zed Books.

Hicks, N. and P. Streeten (1979) 'Indicators of development: the search for a basic needs yardstick', *World Development* 7 (6): 567–80.

Hines, C. (2000) *Localization: A Global Manifesto*, London: Earthscan.

HIPC (Heavily Indebted Poor Countries Initiative) (2010) www.worldbank.org/hipc (accessed 3 August 2010).

Hirschman, A. (1958) *The Strategy of Economic Development*, New Haven: Yale University Press.

Hirst, P., G. Thompson and S. Bromley (2009) *Globalization in Question*, 3rd edition, Cambridge: Polity.

Hodder, R. (2000) *Development Geography*, London: Routledge.

Hogar de Cristo (2010) 'Hogar de Cristo Ecuador', www.hogardecristo.org.ec/ (accessed 1 August 2010).

Holloway, L. and M. Kneafsey (2000) 'Reading the space of the farmers' market: a preliminary investigation from the UK', *Sociologica Ruralis* 40 (3): 285–99.

Hope, Sr, K.R. (2002) 'From crisis to renewal: towards a successful implementation of the New Partnership for Africa's Development', *African Affairs* 101: 387–402.

Hörschelmann, K. (2004) 'The social consequences of transformation', in M. Bradshaw and A. Stenning (eds) *East Central Europe and the Former Soviet Union*, London: Pearson Prentice Hall, pp. 219–46.

Howard, S. (1997) 'Land conflict and Mayangna territorial rights in Nicaragua's Bosawás reserve', *Bulletin of Latin American Research* 17 (1): 17–34.

Howell, J. (1993) *China Opens its Doors: The Politics of Economic Transition*, Hemel Hempstead: Harvester Wheatsheaf.

Howes, D. (ed.) (1996) *Cross-cultural Consumption: Global Markets, Local Realities*, London: Routledge.

Huff, W.G. (1997) *The Economic Growth of Singapore: Trade and Development in the Twentieth Century*, Cambridge: Cambridge University Press.

Huitric, M., C. Folke and N. Kautsky (2002) 'Development and government policy of the shrimp farming industry in Thailand in relation to mangrove ecosystems', *Ecological Economics* 40: 441–55.

Hulme, D. and M. Edwards (1997) *NGOs, States and Donors: Too Close for Comfort?*, Basingstoke: Macmillan in association with Save the Children.

Hunt, D. (1989) *Economic Theories of Development: An Analysis of Competing Paradigms*, London: Harvester Wheatsheaf.

Huntington, E. (1915) *Civilisation and Climate*, New Haven: Yale University Press.

Huq, S., S. Kovats, H. Reid and D. Satterthwaite (2007) 'Editorial: reducing risks to cities from disasters and climate change', *Environment & Urbanization* 19 (1): 3–15.

Hyden, G. (1980) *Beyond Ujamaa in Tanzania*, London: Heinemann.

Inman, P. (2010) 'Finance experts call for "Tobin Tax" on foreign exchange trades', *The Guardian*, 18 July 2010 (available at www.guardian.co.uk).

International Development Planning Review (IDPR) (2010) 'Special issue on Mobility, Migration and Development', *International Development Planning Review* 32 (3–4).

International Labour Organization (ILO) (2006) *The End of Child Labour: Within Reach*, Geneva: ILO (available at: www.ilo.org/public/english/region/ampro/cinterfor/news/rep_ib.pdf).

International Monetary Fund (IMF) (2010a) 'IMF overview' www.imf.org/external/about/overview.htm (accessed 16 July 2010).

International Monetary Fund (IMF) (2010b) 'IMF history' www.imf.org/external/about/history.htm (accessed 16 July 2010).

International Monetary Fund (IMF) (2010c) 'Special Drawing Rights (SDRs) Factsheet', www.imf.org/external/np/exr/facts/pdf/sdr.pdf (accessed 16 July 2010).

International Rivers Network (IRN) (2009) 'China's Three Gorges Dam: a model of the past' (available at: www.internationalrivers.org/files/3Gorges_FINAL.pdf).

Jack, W. and T. Suri (2009) 'Mobile money: the economics of M-PESA', *MIT Working Paper* (available at: http://www.mit.edu/~tavneet/M-PESA.pdf).

Johnston, R. (2009) 'Ethnicity', in D. Gregory, R. Johnston, G. Pratt, M.J. Watts and S. Whatmore (eds) *The Dictionary of Human Geography*, 6th edition, Oxford: Wiley-Blackwell, pp. 214–17.

Jones, P.S. (2000) 'Why is it alright to do development "over there" but not "here"? Changing vocabularies and common strategies of inclusion across "First" and "Third" Worlds', *Area* 32 (2): 237–41.

Jubilee USA (2010) 'Jubilee USA homepage', www.jubileeusa.org/ (accessed 1 August 2010).

Kabeer, N. (2005) 'Gender equality and women's empowerment: a critical analysis of the third millennium development goal 1', *Gender and Development* 13 (1): 13–24.

Kaplinsky, R. (2008) 'What does the rise of Asia do for industrialisation in Sub-Saharan Africa?', *Review of African Political Economy* 35 (115): 7–22.

Kay, C. (2004) 'Rural livelihoods and peasant futures', in R.N. Gwynne and C. Kay (eds) *Latin America Transformed: Globalization and Modernity*, 2nd edition London: Hodder Education, pp. 232–50.

Kelly, P. (2000) *Landscapes of Globalisation*, London: Routledge.

Keynes, J.M. (1936) *The General Theory of Employment, Interest and Money*, London: Macmillan.

Khan Osmani, L.N. (1998) 'The Grameen Bank experiment: empowering women through credit', in H. Afshar (ed.) *Women and Empowerment: Illustrations from the Third World*, Basingstoke: Macmillan, pp. 67–85.

Kilby, P. (2006) 'Accountability for empowerment: dilemmas facing non-governmental organizations', *World Development* 34 (6): 951–63.

Kilmister, A. (2000) 'Socialist models of development', in T. Allen and A. Thomas (eds) *Poverty and Development into the 21st Century*, Oxford: Oxford University Press., pp. 309–24.

Klak, T. (2008) 'World-systems theory: cores, centres, peripheries and semi-peripheries', in V. Desai and R.B. Potter (eds) *The Companion to Development Studies*, 2nd edition, London: Hodder Education, pp. 101–7.

Klein, N. (2000) *No Logo*, London: Flamingo.

Kleine, D. and T. Unwin (2009) 'Technological revolution, evolution and new dependencies: what's new about ICT4D?', *Third World Quarterly* 30 (5): 1045–67.

Koch, D-J. (2008) 'A Paris declaration for international NGOs?', *OECD Development Centre Policy Insights No. 73* (available at www.oecd.org/dev/insights).

Kofoworola, O.F. (2007) 'Recovery and recycling practices in municipal solid waste management in Lagos, Nigeria', *Waste Management* 27: 1139–43.

Kong, L. (1995) 'Music and cultural politics: ideology and resistance in Singapore', *Transactions of the Institute of British Geographers* 20 (4): 447–59.

Kosoy, N., E. Corbera and K. Brown (2008) 'Participation in payments for ecosystem services: case studies from the Lacandon rainforest, Mexico', *Geoforum* 39: 2073–83.

Kragelund, P. (2009) 'Knocking on a wide-open door: Chinese investments in Africa', *Review of African Political Economy* 36 (122): 479–97.

Krause, W. (2009) *Gender and Participation in the Arab Gulf*, London: LSE (available at: www.shebacss.com/docs/soewcsr007–10.pdf).

Kuapa Kokoo (2010) 'Kuapa Kokoo homepage' www.kuapakokoo.com/ (accessed 31 July 2010).

Labour Behind the Label (2010) 'Asian floor wage campaign', www.labourbehindthelabel.org (accessed 3 August 2010).

Lal, D. (1983) *The Poverty of Development Economics*, London: Institute of Economic Affairs.

Lal, D. (1985) 'The misconceptions of "development economics"', *Finance and Development* June: 10–13.

Lancaster, C. (1999) *Aid to Africa: So Much to Do, So Little Done*, Chicago: University of Chicago Press.

Larrain, J. (2004) 'Modernity and identity: cultural change in Latin America', in R.N. Gwynne and C. Kay (eds) *Latin America Transformed: Globalization and Modernity*, 2nd edition, London: Hodder Education, pp. 28–38.

Latham, D. (2010) '100 year-old SA trade agreement to be scrapped', *BusinessDay*, 16 July 2010 (available at: www.businessday.co.za/Articles/Content.aspx?id=115027).

Lazarus, J. (2008) 'Participation in poverty reduction strategy papers: reviewing the past, assessing the present and predicting the future', *Third World Quarterly* 29 (6): 1205–21.

Lewis, D. and N. Kanji (2009) *Non-Governmental Organizations and Development*, London: Routledge.

Lewis, W.A. (1955) *The Theory of Economic Growth*, London: Allen and Unwin.

Lewis, W.A. (1964) 'Economic development with unlimited supplies of labour', in A.N. Agarwala and S.P. Singh (eds) *The Economics of Underdevelopment*, London: Oxford University Press, pp. 400–49 [published originally in 1954 in *The Manchester School of Economic and Social Studies* 22:2].

Leys, C. (1996) *The Rise and Fall of Development Theory*, Oxford: James Currey.

Lindsey, B. (2004) *Grounds for Complaint: 'Fair Trade' and the Coffee Crisis*, London: Adam Smith Institute (available on http://www.adamsmith.org).

Lloyd-Evans, S. (2008) 'Child labour', in V. Desai and R. Potter (eds) *The Companion to Development Studies*, 2nd edition, London: Hodder Arnold, pp. 225–9.

London Citizens (2010) 'London Citizens homepage', www.londoncitizens.org (accessed 3 August 2010).

Loomba, A. (1998) *Colonialism/ Postcolonialism*, London: Routledge.

Luce, S. (2005) 'The role of community involvement in implementing Living Wage ordinances', *Industrial Relations* 44: 32–58.

Lunn, J. (2009) 'The role of religion, spirituality and faith in development: a critical theory approach', *Third World Quarterly* 30 (5): 937–51.

Lynas, M. (2009) 'How do I know China wrecked the Copenhagen deal? I was in the room', *The Guardian* 22 December 2009 (available at: www.guardian.co.uk).

Lynch, K. (2005) *Rural–Urban Interaction in the Developing World*, London: Routledge.

McCarthy, J. and S. Prudham (2004) 'Neoliberal nature and the nature of neoliberalism', *Geoforum* 35 (3): 275–83.

McCormick, D. (2008) 'China & India as Africa's new donors: the impact of aid on development', *Review of African Political Economy* 35 (115): 73–92.

McEwan, C. (2001) 'Postcolonialism, feminism and development: intersections and dilemmas', *Progress in Development Studies* 1 (2): 93–111.

McEwan, C. (2009) *Postcolonialism and Development*, Abingdon: Routledge.

McGranahan, G. (1993) 'Household environmental problems in low-income cities: an overview of problems and prospects for improvement', *Habitat International* 17 (2): 105–21.

McGranahan, G., J. Songsore and M. Kjellén (1999) 'Sustainability, poverty and urban environmental transitions', in D. Satterthwaite (ed.) *The Earthscan Reader in Sustainable Cities*, London: Earthscan, pp. 107–30.

McGranahan, G., D. Balk and B. Anderson (2007) 'The rising tide: assessing the risks of climate change and human settlements in low elevation coastal zones', *Environment & Urbanization* 19 (1): 17–37.

McGregor, D. (2008) 'Climate change and development', in V. Desai and R. Potter (eds) *The Companion to Development Studies*, 2nd edition, London: Hodder Arnold, pp. 282–7.

McGregor, D., D. Simon and D. Thompson (eds) (2006) *The Peri-Urban Interface: Approaches to Sustainable Natural and Human Resource Use*, London: Earthscan.

McIlwaine, C. (1998) 'Civil society and development geography', *Progress in Human Geography* 22 (3): 415–24.

McIlwaine, C. (2002) 'Perspectives on poverty, vulnerability and exclusion', in C. McIlwaine and K. Willis (eds) *Challenges and Change in Middle America: Perspectives on Development in Mexico, Central America and the Caribbean*, London: Pearson, pp. 82–109.

Malthus, T. (1985) *An Essay on the Principle of Population*, London: Penguin Books [published originally in 1798 and 1830].

Manzo, K. (2003) 'Africa in the rise of rights-based development', *Geoforum* 34 (4): 437–56.

Martinussen, J. (1997) *Society, State and Market: A Guide to Competing Theories of Development*, London: Zed Books.

Marx, K. (1909) *Capital*, Volume 1, London: William Glaisher.

Massey, D. (1993) 'Power-geometry and a progressive sense of place', in J. Bird, B. Curtis, T. Putnam, G. Robertson and L. Tickner (eds) *Mapping the Futures: Local Cultures, Global Change*, London: Routledge, pp. 59–69.

Mawdsley, E. (2008) 'Fu Manchu versus Dr Livingstone in the Dark Continent? Representing China, Africa and the West in British broadsheet newspapers', *Political Geography* 27 (5): 509–29.

Mawdsley, E. (2010) 'The Non-DAC donors and the changing landscape of foreign aid: the (in)significance of India's development cooperation with Kenya', *Journal of Eastern African Studies* 4 (2): 361–79.

Maxwell, S. (1999) 'What can we do with a rights-based approach to development?', ODI briefing paper 1999 (3), London: ODI (available on http://www.odi.org.uk).

May, J. (2003) 'Chronic poverty and older people in South Africa', *CPRC Working Paper 25*.

Meadows, D.H., D.L. Meadows, J. Randers and W.W. Behrens III (1972) *The Limits to Growth: A Report for the Club of Rome's Project on the Predicament of Mankind*, London: Pan Books.

Megoran, N. (2009) 'Theocracy', in R. Kitchin and N. Thrift (eds) *International Encyclopedia of Human Geography*, Oxford: Elsevier, pp. 223–8.

Menashri, D. (2001) *Post-Revolutionary Politics in Iran: Religion, Society and Power*, London: Frank Cass.

Menon, R. (2008) 'Kerala's development paradox', *India Together*, 23 March 2008 (available at: www.indiatogether.org/2008/mar/opi-kerala.htm).

Mercer, C. (1999) 'Reconceptualizing state–society relations in Tanzania', *Area* 31 (3): 247–58.

Mercer, C. (2002) 'NGOs, civil society and democratisation: a critical review of the literature', *Progress in Development Studies* 2 (1): 5–22.

Mercer, C., G. Mohan and M. Power (2003) 'Towards a critical political geography of African development', *Geoforum* 34 (4): 419–36.

Mercer, C., B. Page and M. Evans (2008) *Development and the African Diaspora: Place and the Politics of Home*, London: Zed Books.

MERCOSUR (2010) 'Portal official MERCOSUL/MERCOSUR', www.mercosur.int/msweb/Portal%20Intermediario/ (accessed 29 July 2010).

Micklin, P. (2007) 'The Aral Sea disaster', *Annual Review of Earth and Planetary Sciences* 35: 47–72 (available at: www.earth.annualreviews.org).

Middleton, N. (1995) *The Global Casino: An Introduction to Environmental Issues*, London: Arnold.

Milanovic, B. (1998) *Income, Inequality and Poverty During Transition from Planned to Market Economy*, Washington DC: World Bank (available on http://www.worldbank.org).

Milward, B. (2000) 'What is structural adjustment?', in G. Mohan, E. Brown, B. Milward and A.B. Zack-Williams (eds) *Structural Adjustment: Theory, Practice and Impacts*, London: Routledge, pp. 24–38.

Mitchell, K. (2009) 'Communism' in, D. Gregory, R. Johnston, G. Pratt, M.J. Watts and S. Whatmore (eds) *The Dictionary of Human Geography*, 6th edition, Oxford: Wiley-Blackwell, p. 103.

Mohan, G. and J. Holland (2001) 'Human rights and development in Africa: moral intrusion or empowering opportunity?', *Review of African Political Economy* 88: 177–96.

Mohan, G., E. Brown, B. Milward and A.B. Zack-Williams (2000) *Structural Adjustment: Theory, Practice and Impacts*, London: Routledge.

Mohanty, C.T. (1991) 'Under western eyes: feminist scholarship and colonial discourses', in C.T. Mohanty, A. Russo and L. Torres (eds) *Third World Women and the Politics of Feminism*, Bloomington: University of Indiana Press, pp. 51–80.

Molyneux, M. (1987) 'Mobilisation without emancipation: women's interests, the state, and revolution in Nicaragua', *Feminist Studies* 11 (2): 227–54.

Molyneux, M. (2006) 'Mothers at the service of the New Poverty Agenda: Progresa/Oportunidades, Mexico's conditional transfer programme', *Social Policy & Administration* 40 (4): 425–49.

Momsen, J. (2010) *Gender and Development*, 2nd edition, London: Routledge.

Monteiro, O. (1999) 'Governance and decentralization', in B. Ferraz and B. Munslow (eds) *Sustainable Development in Mozambique*, Oxford: James Currey, pp. 29–45.

Morse, S. (2004) *Indices and Indicators in Development: An Unhealthy Obsession with Numbers*, London: Earthscan.

Moser, C. (1993) *Gender Planning and Development: Theory, Practice and Training*, London: Routledge.

Moser, C. and C. McIlwaine (1999) 'Participatory urban appraisal and its application for research on violence', *Environment and Urbanization* 11 (2): 203–26.

Mosse, D. (2001) '"People's knowledge", participation and patronage: operations and representations in rural development', in B. Cooke and

U. Kothari (eds) *Participation: The New Tyranny?*, London: Zed Books, pp. 16–35.

Moyo, D. (2010) *Dead Aid: Why Aid Is Not Working and How There Is Another Way for Africa*, London: Penguin.

Mullings, B. (2009) 'Neoliberalization, social reproduction and the limits to labour in Jamaica', *Singapore Journal of Tropical Geography* 30 (2): 174–88.

Multilateral Investment Guarantee Agency (MIGA) (2010) 'About MIGA' www.miga.org/about/index_sv.cfm?stid=1736 (accessed 16 July 2010).

Myrdal, G. (1957) *Economic Theory and Underdeveloped Regions*, London: Gerald Duckworth.

Myrdal, G. (1970) *The Challenge of World Poverty*, London: Allen Lane.

Naidu, S. (2008) 'India's growing African strategy', *Review of African Political Economy* 35 (115): 116–28.

Nederveen Pieterse, J. (2000) 'After post-development', *Third World Quarterly* 21 (2): 175–91.

New Internationalist (1993) Kerala, *New Internationalist* 241 (available on http://www.newint.org).

NTNC (National Trust for Nature Conservation) (2010) 'Annapurna Conservation Area Project', http://dev.yipl.com.np/ntnc/projects (accessed 31 July 2010).

Nursey-Bray, M. (2009) '*A Guugu Yimmithir Bam Wii: Ngawiya and Girrbithi*: Hunting, planning and management along the Great Barrier Reef, Australia', *Geoforum* 40 (3): 442–53.

OECD (2008) 'Is it ODA?' *OECD Factsheet November 2008* (available at www.oecd.org/dataoecd/21/21/34086975.pdf).

OECD (2010) 'OECD homepage', http://www.oecd.org (accessed 30 July 2010).

Office of the UN High Commission for Human Rights (OHCHR) (2010) 'Convention on the Rights of the Child', www2.ohchr.org/english/law/crc.htm (accessed 1 August 2010).

Ogborn, M. (2005) 'Modernity and modernization', in P. Cloke, P. Crang and M. Goodwin (eds) *Introducing Human Geographies*, 2nd edition, London: Arnold, pp. 339–49.

O'Keefe, P. (2007) *People with Disabilities in India: From Commitments to Outcome*, Washington: World Bank Human Development Unit South Asia Region.

O'Riordan, T. (1981) *Environmentalism*, 2nd edition, London: Pion.

Owusu, F. (2003) 'Pragmatism and the gradual shift from dependency to neoliberalism: the World Bank, African leaders and development policy in Africa', *World Development* 31 (10): 1655–72.

Páez-Osuna, F. (2001) 'The environmental impact of shrimp aquaculture: causes, effects and mitigating alternatives', *Environmental Management* 28 (1): 131–40.

Page, J. (1994) 'The East Asian Miracle: an introduction', *World Development* 22 (4): 615–25.

Pagiola, S., A. Arcenas and G. Platais (2005) 'Can payments for environmental services help to reduce poverty? An exploration of the issues and evidence to date from Latin America', *World Development* 33 (2): 237–53.

Panizza, F. (2009) *Contemporary Latin America: Development and Democracy beyond the Washington Consensus*, London: Zed Books.

Parnwell, M. (2003) 'Consulting the poor in Thailand: enlightenment or delusion?', *Progress in Development Studies* 3 (2): 99–112.

Parreñas, R.S. (2001) *Servants of Globalization: Women, Migration and Domestic Work*, Stanford: Stanford University Press.

Parreñas, R.S. (2005) *Children of Global Migration: Transnational Families and Gendered Woes*, Stanford: Stanford University Press.

Parsons, T. (1951) *The Social System*, London: Routledge & Kegan Paul.

Parsons, T. (1966) *Societies: Evolutionary and Comparative Perspectives*, Englewood Cliffs: Prentice Hall.

Pattison, V. (2008) 'Neoliberalization and its discontents: the experience of working poverty in Manchester', in A. Smith, A. Stenning and K. Willis (eds) *Social Justice and Neoliberalism: Global Perspectives*, London: Zed Books, pp. 90–113.

Peet, R. (2007) *Geography of Power*, London: Zed Books.

Peet, R. and E. Hartwick (2009) *Theories of Development*, 2nd edition, London: Guilford Press.

Pelling, M. (2002) 'Assessing urban vulnerability and social adaptation to risk: evidence from Santo Domingo', *International Development Planning Review* 24 (1): 59–76.

Pepper, D. (1996) *Modern Environmentalism: An Introduction*, London: Routledge.

Perry, M., L. Kong and B. Yeoh (1997) *Singapore: A Developmental City State*, Chichester: Wiley.

Pobocik, M. and C. Butalla (1998) 'Development in Nepal: the Annapurna Conservation Area Project', in C.M. Hall and A.A. Lew (eds) *Sustainable Tourism: A Geographical Perspective*, Harlow: Longman, pp. 159–72.

Poon, J. and M. Perry (1999) 'The Asian economic "flu": a geography of crisis', *Professional Geographer* 51 (2): 184–96.

Porter, G. (2003) 'NGOs and poverty reduction in a globalizing world: perspectives from Ghana', *Progress in Development Studies* 3 (2): 131–45.

Potter, R.B., T. Binns, J.A. Elliott and D. Smith (2008) *Geographies of Development: An Introduction to Development Studies*, 3rd edition, London: Pearson Education.

Power, M. (2003) *Rethinking Development Geographies*, London: Routledge.

Practical Action (2010) 'Small-scale wind power', www.practicalaction.org.uk/energy-advocacy/access-wind-sri-lanka (accessed 30 July 2010).

Prebisch, R. (1959) 'International trade and payments in an era of coexistence: commercial policy in the underdeveloped countries', *American Economic Review* 49 (2): 251–273.

Preston, P.W. (1996) *Development Theory: An Introduction*, Oxford: Blackwell.

Purvis, A. (2003) 'The tribe that survives on chocolate', *The Observer Food Monthly* 32: 22–32 (available on http://observer.guardian.co.uk/).

Radcliffe, S. (2005) 'Rethinking development', in P. Cloke, P. Crang and M. Goodwin (eds) *Introducing Human Geographies*, 2nd edition, London: Hodder Arnold, pp. 200–09.

Radcliffe, S. (2001) 'Development, the state and transnational political connections: state and subject formations in Latin America', *Global Networks* 1 (1): 19–36.

Radcliffe, S. (2006) 'Culture in development thinking: geographies, actors and paradigms', in S.A. Radcliffe (ed.) *Culture and Development in a Globalizing World: Geographies, Actors and Paradigms*, London: Routledge, pp. 1–29.

Raffer, K. (1998) 'The Tobin Tax: reviving a discussion', *World Development* 26 (3): 529–38.

Raghuram, P. (1999) 'Religion and development', in T. Skelton and T. Allen (eds) *Culture and Global Change*, London: Routledge, pp. 232–9.

Rahnema, M. with V. Bawtree (eds) (1997) *The Post-Development Reader*, London: Zed Books.

Reid, M. (2007) *Forgotten Continent: The Battle for Latin America's Soul*, London: Yale University Press.

Ridge, T. (2002) *Childhood Poverty and Social Exclusion: From a Child's Perspective*, Bristol: Policy Press.

Rigg, J. (2003) *Southeast Asia*, 2nd edition, London: Routledge.

Rist, G. (2008) *The History of Development: From Western Origins to Global Faith*, 3rd edition, London: Zed Books.

Robson, E. (2000) 'Invisible carers: young people in Zimbabwe's home-based healthcare', *Area* 32 (1): 59–69.

Robson, E. (2004) 'Hidden child workers: young carers in Zimbabwe', *Antipode* 36 (2): 227–48.

Robson, E. and N. Ansell (2000) 'Young carers in Southern Africa: exploring stories from Zimbabwean secondary school students', in S.L. Holloway and G. Valentine (eds) *Children's Geographies: Playing, Living, Learning*, London: Routledge, pp. 174–93.

Rodney, W. (1981) *How Europe Underdeveloped Africa*, Washington DC: Howard University Press [originally published by Tanzania Publishing House in 1972].

Rostow, W.W. (1960) *The Stages of Economic Growth: A Non-Communist Manifesto*, Cambridge: Cambridge University Press.

Rowlands, I.H. (2001) 'The Kyoto Protocol's "Clean Development Mechanism": a sustainability assessment', *Third World Quarterly* 22 (5): 795–811.

Rowlands, J. (1997) *Questioning Empowerment: Working with Women in Honduras*, Oxford: Oxfam.

Rowlands, J. (1998) 'A word of our time, but what does it mean? Empowerment in the discourse and practice of development', in H. Afshar (ed.) *Women and Empowerment: Illustrations from the Third World*, Basingstoke: Macmillan, pp. 11–34.

Roxborough, I. (1979) *Theories of Underdevelopment*, Basingstoke: Macmillan.

Ruiz, N.G. (2008) 'Managing migration: lessons from the Philippines', *World Bank Migration and Development Brief 6* (available at: http://siteresources. worldbank.org/INTPROSPECTS/Resources/334934–1110315015165/ MD_Brief6.pdf).

Sachs, W. (ed.) (1992) *The Development Dictionary: A Guide to Knowledge as Power*, London: Zed Books.

SACU (2010) 'SACU homepage', www.sacu.int/index.php (accessed 29 July 2010).

Said, E. (1991) *Orientalism: Western Conceptions of the Orient*, London: Penguin Books [originally published by Routledge & Kegan Paul in 1978].

Salehi-Isfahani, D. (2009) 'Poverty, inequality and populist politics in Iran', *Journal of Economic Inequality* 7 (1): 5–28.

Samers, M. (2010) *Migration*, London: Routledge.

Schech, S. and J. Haggis (2000) *Culture and Development: A Critical Introduction*, Oxford: Blackwell.

Schumacher, E.F. (1974) *Small is Beautiful: A Study of Economics as if People Mattered*, London: Abacus [originally published by Blond and Briggs, London in 1974].

Schuurman, F.J. (1993) 'Introduction: development theory in the 1990s' in F.J. Schuurman (ed.) *Beyond the Impasse: New Directions in Development Theory*, London: Zed Books, pp. 1–48.

Semple, E.C. (1911) *Influences of Geographic Environment on the Basis of Ratzel's System of Anthropo-Geography*, New York: Russell & Russell.

Sen, A. (1999) *Development as Freedom*, Oxford: Oxford University Press.

Shannon, R. (2009) 'Playing with principles in an era of securitized aid: negotiating humanitarian space in post-9/11 Afghanistan', *Progress in Development Studies* 9 (1): 15–36.

Sharma, J. (2008) 'The language of rights' in A. Cornwall, S.S. Corrêa and S. Jolly (eds) *Development with a Body: Sexuality, Human Rights & Development*, London: Zed Books, pp. 67–76.

Sharp, J. (2009) *Geographies of Postcolonialism*, London: Sage.

Sharpley, R. and D.J. Telfer (2008) *Tourism and Development*, London: Routledge.

Sheppard, E., P.W. Porter, D.R. Faust and R. Nagar (2009) *A World of Difference: Encountering and Contesting Development*, 2nd edition, New York: Guilford Press.

Shiva, V. (1991) *The Violence of the Green Revolution*, London: Zed Books.

Shrestha, N. (1995) 'Becoming a development category', in J. Crush (ed.) *Power of Development*, London: Routledge, pp. 266–77.

Sidaway, J. (2008) 'Post-development', in V. Desai and R. Potter (eds) *The Companion to Development Studies*, 2nd edition, London: Hodder Arnold, pp. 16–20.

Simon, D. (1994) 'Imperialism: the struggle for Africa', in T. Unwin (ed.) *Atlas of World Development*, Chichester: John Wiley & Sons, pp. 19–21.

Simon, D. (1995) 'Debt, democracy and development: Sub-Saharan Africa in the 1990s', in D. Simon, W. Van Spengen, C. Dixon and A. Närman (eds) *Structurally Adjusted Africa: Poverty, Debt and Basic Needs*, London: Pluto Press, pp. 17–44.

Simon, D. (1998) 'Rethinking (post)modernism, postcolonialism, and posttradtionalism: South–North perspectives', *Environment and Planning D: Society and Space* 16 (2): 219–45.

Simon, D. (2008) 'Neoliberalism, structural adjustment and poverty reduction strategies', in V. Desai and R. Potter (eds) *The Companion to Development Studies*, 2nd edition, London: Hodder Education, pp. 86–92.

Simon, D., W. Van Spengen, C. Dixon and A. Närman (eds) (1995) *Structurally Adjusted Africa: Poverty, Debt and Basic Needs*, London: Pluto Press.

Simon, J. (1997) *Endangered Mexico: An Environment on the Edge*, London: Latin American Bureau.

Six, C. (2009) 'The rise of postcolonial states as donors: a challenge to the development paradigm?', *Third World Quarterly* 30 (6): 1103–21.

Skelton, T. (2008) 'Children, young people, UNICEF and participation', in S. Aitkin, R. Lund and A.T. Kjørholt (eds) *Global Childhoods: Globalization, Development and Young People*, London: Routledge, pp. 165–81.

Skidmore, T.E. and P.H. Smith (2004) *Modern Latin America*, 6th edition, Oxford: Oxford University Press.

Smith, A. (1950) *An Inquiry into the Nature and Causes of the Wealth of Nations*, London: Methuen [edited by Edwin Cannon].

Smith, A. and A. Padulla (1996) *Sex and Revolution: Women in Socialist Cuba*, Oxford: Oxford University Press.

Sparks, D.L. (2004) 'Economic trends in Africa south of the Sahara, 2003', in *Africa South of the Sahara 2004*, 33rd edition, London: Europa Publications, pp. 12–21.

Spencer, H. (1975) *The Principles of Sociology*, Westport: Greenwood Press [originally published by Williams and Norgate, London in 1876–9].

Spoor, M. (1998) 'The Aral Sea Basin crisis: transition and environment in former Soviet Central Asia', *Development and Change* 29: 409–35.

Stavenhagen, R. (1986) 'Ethnodevelopment: a neglected dimension in development thinking', in R. Anthorpe and A. Kráhl (eds) *Development Studies: Critique and Renewal*, Leiden: E.J. Brill, pp. 71–94.

Stavenhagen, R. (1996) *Ethnic Conflicts and the Nation-State*, Basingstoke: Macmillan in association with UNRISD.

Stavenhagen, R. (2003) 'Needs, rights and social development', *Overarching Concerns Paper*, 2, July, Geneva: UNRISD (available on http://www.unrisd.org).

Stewart, F. (1995) *Adjustment and Poverty: Options and Choices*, London: Routledge.

Stiglitz, J. and A. Charlton (2005) *Fair Trade for All: How Trade can Promote Development*, Oxford: Oxford University Press.

Stock, R. (2004) *Africa South of the Sahara: A Geographical Interpretation*, 2nd edition, London: Guilford Press.

Stoneman, C. (2004) 'Structural adjustment in eastern and southern Africa', in D. Potts and T. Bowyer-Bower (eds) *Eastern and Southern Africa: Development Challenges in a Volatile Region*, London: Pearson, pp. 58–88.

Sutton, K. and S.E. Zaimeche (2002) 'The collapse of state socialism in the socialist third world', in V. Desai and R. Potter (eds) *The Companion to Development Studies*, London: Arnold, pp. 20–6.

Sweetman, C. (2000) 'Editorial' in issue on 'Gender and Lifecycles', *Gender and Development* 8 (2): 2–8.

Sylvester, C. (1999) 'Development studies and postcolonial studies: disparate tales of the "Third World"', *Third World Quarterly* 20 (4): 703–21.

Szeftel, M. (2006) 'Sir William Arthur Lewis', in D. Simon (ed.) *Fifty Key Thinkers on Development*, Abingdon: Routledge, pp. 144–9.

Tacoli, C. (ed.) (2006) *The Earthscan Reader in Rural–Urban Linkages*, London: Earthscan.

Tang, S. and G. Bloom (2000) 'Decentralizing rural health services: a case study in China', *International Journal of Health Planning and Management* 15: 189–200.

Taylor, P and P. Bain (2005) '"India calling to the far away towns": the call centre labour process and globalization', *Work, Employment and Society* 19 (2): 261–82.

Thomas, A. and T. Allen (2000) 'Agencies of development', in T. Allen and A. Thomas (eds) *Poverty and Development into the 21st Century*, Oxford: Oxford University Press, pp. 189–216.

Thompson, C.B. (2007) 'Green revolution or rainbow evolution?' *Foreign Policy in Focus*, 17 July 2007 (available at: www.fpif.org/articles/africa_green_revolution_or_rainbow_evolution).

Thorpe, A. and E. Bennett (2002) 'Sowing the seeds of modernity: the insertion of agriculture into the global market', in C. McIlwaine and K. Willis (eds) *Challenges and Change in Middle America: Perspectives on Development in Mexico, Central America and the Caribbean*, London: Pearson, pp. 159–90.

Tickle, L. (2004) 'Raising the bar', *Developments* 25: 4–8 (available on http://www.developments.org.uk).

Tobin, J. (1994) 'A tax on international currency transactions', in UNDP *Human Development Report 1994*, Oxford: Oxford University Press.

Todaro, M.P. (2000) *Economic Development in the Third World*, 7th edition, London: Longman.

Tomalin, E. (2006) 'Religion and a rights-based approach to development', *Progress in Development Studies* 6 (2): 93–108.

Tomlinson, J. (1999) *Globalization and Culture*, Cambridge: Polity.

Tourism Concern (2010) 'Trekking wrongs: porters rights' http://www.tourismconcern.org.uk/campaigns/campaigns_porters.htm (accessed 31 July 2010).

Townsend, J., E. Zapata, J. Rowlands, P. Alberti and M. Mercado (1999) *Women and Power: Fighting Patriarchies and Poverty*, London: Zed Books.

Toye, J. (1993) *Dilemmas of Development*, 2nd edition, Oxford: Blackwell.

Tsai, K.S. (2006) 'Debating decentralized development: a reconsideration of the Wenzhou and Kerala models', *Indian Journal of Economic and Business*, (special issue: India and China): 46–67

Tullos, D. (2009) 'Assessing the influence of environmental impact assessments on science and policy: an analysis of the Three Gorges Project', *Journal of Environmental Management* 9: S208–23.

UNDP (1994) *Human Development Report 1994*, Oxford: Oxford University Press.

UNDP (1995) *Human Development Report 1995*, Oxford: Oxford University Press.

UNDP (2002) *Human Development Report 2002*, Oxford: Oxford University Press.

UNDP (2003) *Human Development Report 2003*, Oxford: Oxford University Press.

UNDP (2007) *Human Development Report 2007/8,* Basingstoke: Palgrave Macmillan (available at www.undp.org).

UNDP (2009) *Human Development Report 2009,* Basingstoke: Palgrave Macmillan. (available at www.undp.org).

UNDP (2010a) *Human Development Report 2010*, Basingstoke: Palgrave Macmillan. (available at www.undp.org).

UNDP (2010b) 'Tracking global progress', www.undp.org/mdg/progress.shtml (accessed 10 August 2010).

UNDP and the Institute of National Planning, Egypt (2010) *Egypt Human Development Report 2010* (available at: http://hdr.undp.org/en/reports/ nationalreports/arabstates/egypt/Egypt_2010_en.pdf).

UN-Habitat (2008) *The State of African Cities 2008*, Nairobi: UN-HABITAT (available at www.unhabitat.org).

UN-Habitat (2010) *State of the World's Cities, 2010/2011: Bridging the Urban Divide*, Nairobi: UN-HABITAT.

UNICEF (2003) 'UN Convention on the Rights of the Child', http://www.unicef.org/crc/fulltext.htm (accessed 18 November 2003).

Unwin, T. (ed.) (2009) *ICT4D*, Cambridge: Cambridge University Press.

USAID (2010) 'Where does USAID's money go?' www.usaid.gov/policy/budget/money/ (accessed 28 July 2010).

Vakil, A.C. (1997) 'Confronting the classification problem: toward a taxonomy of NGOs', *World Development* 25 (12): 2057–70.

Van Rooy, A. (1998) *Civil Society and the Aid Industry*, London: Earthscan.

Van Rooy, A. (2008) 'Strengthening civil society in developing countries', in V. Desai and R.B. Potter (eds) *The Companion to Development Studies*, 2nd edition, London: Hodder Education, pp. 520–5.

Véron, R. (2001) 'The "new" Kerala model: lessons for sustainable development', *World Development* 29 (4): 601–17.

Vertovec, S. (1999) 'Conceiving and researching transnationalism', *Ethnic and Racial Studies* 22 (2): 447–62.

Vertovec, S. (2009) *Transnationalism*, London: Routledge.

Vickers, J. (1991) *Women and the World Economic Crisis*, London: Zed Books.

Vidal, J., A. Stratton and S. Goldenberg (2009) 'Low targets, goals dropped: Copenhagen ends in failure', *The Guardian*, 19 December 2009 (available at: www.guardian.co.uk).

Vujakovic, P. (1989) 'Mapping for world development', *Geography* 74 (2): 97–105.

Wallerstein, I. (1974) *The Modern World System*, New York: Academic Press.

War on Want (2010) 'The Robin Hood Tax', www.waronwant.org/campaigns/financial-crisis/the-robin-hood-tax (accessed 10 August 2010).

Watts, J. (2009) 'What was agreed at Copenhagen – and what was left out?' *The Guardian*, 19 December 2009 (available at: www.guardian.co.uk).

Watts, M.J. (2009) 'Capitalism', in D. Gregory, R. Johnston, G. Pratt, M.J. Watts and S. Whatmore (eds) *The Dictionary of Human Geography*, 6th edition, Oxford: Wiley-Blackwell, pp. 58–64.

Weber, M. (1930) *The Protestant Ethic and the Spirit of Capitalism*, trans. T. Parsons, London: Allen & Unwin.

Webster, A. (1990) *Introduction to the Sociology of Development*, 2nd edition, Basingstoke: Macmillan.

Wei Y. and L.J.C. Ma (1996) 'Changing patterns of spatial inequality in China, 1952–1990', *Third World Planning Review* 18 (2): 177–91.

Weiss, T.G. (2000) 'Governance, good governance and global governance: conceptual and actual challenges', *Third World Quarterly* 21 (5): 795–814.

Welsh, J. (2000) 'Organizing the scale of labor regulation in the United States: service-sector activism in the city', *Environment and Planning A* 32: 1593–610.

White, B. (1996) 'Globalization and the child labour problem', *Journal of International Development* 8 (6): 829–39.

White, H. (2008) 'The measurement of poverty', in V. Desai and R. Potter (eds) *The Companion to Development Studies*, 2nd edition, London: Hodder Education, pp. 25–30.

Williams, G. (2004) 'Reforming Africa: continuities and changes', in *Africa South of the Sahara 2004*, 33rd edition, London: Europa Publications, pp. 3–11.

Williams, G. and C. McIlwaine (2003) 'Entanglements of participation: the theory and practices of attacking poverty', *Progress in Development Studies* 3 (2): 93–7.

Williams, G., P. Meth and K. Willis (2009) *Geographies of Developing Areas: The Global South in a Changing World*, London: Routledge.

Williamson, J. (2000) 'What should the World Bank think about the Washington Consensus?', *The World Bank Research Observer* 15 (2): 251–64 (available on http://www.worldbank.org).

Willis, K. (2002) 'Open for business: strategies for economic diversification', in C. McIlwaine and K.Willis (eds) *Challenges and Change in Middle America: Perspectives on Development in Mexico, Central America and the Caribbean*, London: Pearson, pp. 136–58.

Willis, K. (2006) 'Norman Borlaug', in D. Simon (ed.) *Fifty Key Thinkers in Development*, London: Routledge, pp. 45–50.

Willis, K. and S. Khan (2009) 'Health reform in Latin America and Africa: decentralisation, participation and inequalities', *Third World Quarterly* 30 (5): 991–1005.

Willis, K., A. Smith and A. Stenning (2008) 'Introduction: social justice and neoliberalism', in A. Smith, A. Stenning and K. Willis (eds) *Social Justice and Neoliberalism: Global Perspectives*, London: Zed Books, pp. 1–15.

Wills, J. (2009) 'The living wage', *Soundings: A Journal of Politics and Culture*, 42: 33–46.

Wills, J. (2010) 'Researching London's Living Wage campaign', www.geog.qmul.ac.uk/livingwage/index.html (accessed 3 August 2010).

Wills, J., K. Datta, Y. Evans, J. Herbert, J. May and C. McIlwaine (2009) *Global Cities at Work: New Migrant Divisions of Labour*, London: Pluto Press.

Wisner, B., P.M. Blaikie, T. Cannon and I. Davis (2003) *At Risk: Natural Hazards, People's Vulnerability and Disasters*, 2nd edition, London: Routledge.

Woodhouse, P. (2000) 'Environmental degradation and sustainability', in T. Allen and A. Thomas (eds) *Poverty and Development into the 21st Century*, Oxford: Oxford University Press, pp. 141–62.

Wood, D. (2010) *Rethinking the Power of Maps*, New York: Guilford Press.

World Bank (1983) *World Development Report 1983*, Oxford: Oxford University Press.

World Bank (1984) *World Development Report 1984*, Oxford: Oxford University Press.

World Bank (1993) *The East Asian Miracle*, Oxford: Oxford University Press.

World Bank (1994) *World Development Report 1994*, Oxford: Oxford University Press.

World Bank (2003) *World Development Report 2003*, Oxford: Oxford University Press.

World Bank (2008) *2008 World Development Indicators Poverty Data: A Supplement to World Development Indicators 2008*, Washington DC: World Bank. (available at www.worldbank.org).

World Bank (2010a) 'IBRD background', http://go.worldbank.org/D6IEM83I10 (accessed 16 July 2010).

World Bank (2010b) 'Poverty reduction strategies', http://go.worldbank.org/ FXXJK3VEW0 (accessed 16 July 2010).

World Bank (2010c) 'What is IDA?', http://go.worldbank.org/ZRAOR8IWW0 (accessed 16 July 2010).

World Bank (2010d) 'What is social capital?', http://go.worldbank.org/K4LUMW43B0 (accessed 28 July 2010).

World Bank (2010e) *World Development Report 2010*, Washington: World Bank.

World Bank (2010f) 'Governance and anti-corruption', www.worldbank.org/wbi/governance (accessed 4 August 2010).

World Commission on Environment and Development (WCED) (1987) *Our Common Future*, Oxford: Oxford University Press.

World Trade Organization (WTO) (2009) *International Trade Statistics 2009*, Geneva: WTO (available at: www.wto.org).

World Trade Organization (WTO) (2010) 'World Trade Organization homepage', www.wto.org (accessed 9 August 2010).

World Travel and Tourism Council (WTTC) (2010) 'Tourism economic data search tool', www.wttc.org/eng/Tourism_Research/Economic_Data_Search_Tool/ (accessed 31 July 2010).

Worsley, P. (1990) *Marx and Marxism*, London: Routledge.

Worsley, P. (1999) 'Culture and development theory', in T. Skelton and T. Allen (eds) *Culture and Global Change*, London: Routledge, pp. 13–21.

Worster, D. (2004) *Dust Bowl: The Southern Plains in the 1930s*, Oxford: Oxford University Press [originally published 1979].

Wratten, E. (1995) 'Conceptualizing urban poverty', *Environment and Urbanization* 7 (11): 11–36.

Wu, C.-T. (1987) 'Chinese socialism and uneven development', in D. Forbes and N. Thrift (eds) *The Socialist Third World: Urban Development and Territorial Planning*, Oxford: Blackwell, pp. 53–97.

Yapa, L. (1993) 'What are improved seeds? An epistemology of the Green Revolution', *Economic Geography* 69 (3): 254–73.

Yin, S.L. and B. Pearson (n.d.) 'Clean development or development jeopardy? An exploration of risks associated with FDI aspects of the CDM' (available on http://www.cdmwatch.org).

Young, C. (1982) *Ideology and Development in Africa*, New Haven: Yale University Press.

Zeppel, H. (1998) 'Land and culture: sustainable tourism and indigenous peoples', in C.M. Hall and A.A. Lew (eds) *Sustainable Tourism: A Geographical Perspective*, Harlow: Longman, pp. 60–74.

Zhang, L. and S.S. Han (2009) 'Regional disparities in China's urbanisation: an examination of trends 1982–2007', *International Development Planning Review* 31 (4): 355–76.

Index

Note: where an entry appears in a figure, table or box, the page reference appears in **bold**.